sky dance of the wo

A BUR OAK BOOK

Holly Carver, series editor

sky dance of the woodcock   by greg hoch

the habits
and habitats
of a strange
little bird

university of iowa press   iowa city

University of Iowa Press, Iowa City 52242

Copyright © 2019 by the University of Iowa Press

www.uipress.uiowa.edu

Printed in the United States of America

Design by Omega Clay

The University of Iowa Press is a member of Green Press Initiative and is committed to preserving natural resources.

Printed on acid-free paper

Library of Congress Cataloging-in-Publication Data

Names: Hoch, Greg, 1971– author.

Title: Sky dance of the Woodcock : the habits and habitats of a strange little bird / by Greg Hoch.

Description: Iowa City : University of Iowa Press, [2019] | Series: A bur oak book | Includes bibliographical references and index.

Identifiers: LCCN 2018030736 (print) | LCCN 2018032112 (ebook) | ISBN 978-1-60938-628-3 | ISBN 978-1-60938-627-6 (paperback : alk. paper)

Subjects: LCSH: Woodcock, American.

Classification: LCC QL696.C48 (ebook) | LCC QL696.C48 H63 2019 (print) | DDC 598.3/3—dc23

LC record available at https://lccn.loc.gov/2018030736

to Kevin

and

to Joyce and Annie

I owned my farm for two years before becoming aware that the sky dance is to be seen in my wood lot, each evening in April and May. Once discovered, my family and I are reluctant to miss even a single performance.
—Aldo Leopold, *A Sand County Almanac*

More people should learn the sky dance; we cannot conserve what we do not know exists.—Aldo Leopold, *For the Health of the Land*

contents

acknowledgments

Jim and Pat, Sarah, Dave, Commissioner Landwehr, thanks for all the opportunities you've given me over the last couple years. Paul and Bryan, thanks for your leadership within the wildlife section. Bob, Ricky, Leslie, Steve, Carrol, Grant, Katie, Kelly, Linda, Jodi, Kristel, Steph, Deb, Jen, Jay, Rick, and everyone else on the second floor, I appreciate your encouragement and camaraderie. Jessica, MB, and Scott, I thank you for all your assistance with our program.

I came to woodcock through research with golden-winged warblers that my students, especially Tony Hewitt, and I conducted at Tamarac National Wildlife Refuge in northwestern Minnesota. Wayne Brininger, refuge biologist at Tamarac, as well as Earl Johnson and Donna Dustin with the Minnesota Department of Natural Resources guided most of my early initiation into the world of the timberdoodle. Just as important, I should thank Earl's and Donna's dogs for pointing me through the habitat, literally. My conversations with Tom Cooper, Kyle Daly, and David Andersen over the years have only enriched and increased my interest in these birds. More generally, all science is built on the foundation of those who went before. Judging from the heavy use of quotations in this book, it could not have been written without the scores of researchers, poets, and nature writers who have focused on this little bird in the past.

I also wish to thank the staff at the University of Iowa Press for their encouragement and thoughtful comments on earlier drafts of this manuscript. I especially thank Holly Carver, Ranjit Arab, and Meredith Stabel.

Dave Krohne was my undergraduate professor and is still my professor. I continue to rely on his advice on a daily to weekly basis. A lot of this book talks about landscapes. My graduate studies were guided by John Briggs, who taught me to look at patterns and processes of landscapes. Together, these two men largely shaped who I am as a scientist.

Of course, Mom and Dad set the foundation that they could build on. My parents gave me just the right amount of guidance and freedom as a kid so that I could explore parts of the natural world alone. Today, Deanna often accompanies me on excursions into field and forest.

To save one of the best for last, Tim, thank you for both your friendship and your mentorship.

And finally, without dogs in our lives, we wouldn't have lives. Every day is an adventure in woods and fields with dogs like Boomer and Ding.

preface

In *A Sand County Almanac*, Aldo Leopold writes that while "I love trees, I am in love with pines." In one of the most famous letters in American history, Thomas Jefferson outlines an argument between his head and his heart. While the aspen, birch, alders, tamaracks, ruffed grouse, golden-winged warblers, and other components of the North Woods have my head, woodcock lay sole—indeed, soul—claim to my heart.

Wildlife conservation can be boiled down to two basic elements, the rational and the irrational. First, we have to rationally and intellectually learn about and understand which species or habitat we are trying to save, conserve, or restore.

> More people should learn the sky dance; we cannot conserve what we do not know exists. (Leopold 1999)

Second, we need to have some irrational element of love for the species or habitat. This must be matched with the drive, passion, and excitement to do the necessary work.

> But it's damn tough to spend ten springs watching a jaunty little cock bird performing aerobatics that outdo the Blue Angels, seeing woodcock behave with cocky perversity, and not read a bit of impish fun into what they do. That's human nature. We tend to worry about the things we love, whether the worry is justified or not. (Vance 1981)

> . . . an odd, reclusive bird who delights those who see him and fascinate those who study him. (de la Valdène 1985)

> And while it flies in the face of logic that so small a creature could occupy so much real estate in a hunter's heart, to those who fancy him, there's no need to elaborate further. (Osborn 2016)

Few species can capture the imagination as well as the woodcock. This book relies heavily on the abundant historic and modern literature on this species. As the reader will see, woodcock seem to bring out the poet in almost anyone who writes about them.

This book has three goals. The first is to introduce you to the world of the American woodcock and get you outside some March or April evening to watch the sky dance. If you wish, you can also grab an old side-by-side shotgun and follow a good dog through the October aspen and birch. The second is to help you understand the decisions needed to manage habitats specifically for woodcock and more generally for all wildlife species. Last, I hope this book shows everyone that while deforestation has an ugly history in some places and at some times, an ax or a saw can be used carefully, thoughtfully, and constructively. Many species of wildlife can benefit from the same practices that also sustain rural forestry economies.

This book will focus more on the ecology and management of woodcock and less on modern hunting. That niche has already been filled by dozens of books on upland hunting generally and woodcock hunting specifically. However, many of the quotes in this book and many of the references in the bibliography cover a number of hunting sources.

I recently had the pleasure of spending a couple days with Stan Temple, who held Leopold's position at the University of Wisconsin before his retirement and is now a senior fellow at the Aldo Leopold Foundation. He stressed the emphasis that Leopold placed on the management of private lands. While state and federal governments do own large tracts of state and national forests, woodcock like many species will always be heavily dependent on the decisions that private landowners make.

It's not enough to just have science or just have art in our lives. The woodcock is one of the best ways to see how those two meet to tell a more complete story. Without both, any story about this little bird would be only partially told.

sky dance of the woodcock

introduction

Everyone knows when bluebirds and warblers return in the springtime. Their bright feathers and loud songs alert everyone to their presence. Even the brown sparrows make themselves known by their songs and the crowds they form at backyard bird feeders. Waterfowl suddenly appear on lakes and wetlands. Flocks of shorebirds materialize almost magically on beaches and mudflats.

Woodcock arrive by night, quietly sequester themselves in thickets of vegetation before morning's first light, and are rarely seen by any of us. They live in out-of-the-way places in habitats that few people would choose to explore. They are such masters of camouflage that even when we are in their presence, we often don't know it. Woodcock are possibly one of the least-known and least-recognized birds in the eastern United States.

Something light as the whisper of a ghost lifted from the dense bottom growth, spiraled up like an animated maple leaf. An etching of an old master. Only more like a motion picture etching, if you can imagine that. (MacQuarrie 1940)

... the strange, eerie winged ghost of the forest lowlands. (Betten 1940)

It seems a spirit unfettered for a time, transcending its ordinary days, attaining a superlative moment. There is something symbolic about its flight. We too feel a lift in spirits as we follow it with our eyes as it speeds across the sky. (Teale 1974)

Woodcock are anatomical oddities. They have enormous eyes, so large that the brain is rotated backward in the skull to make room for them. They are one of the few birds with a flexible beak that can open to grasp worms and other prey while probing the soil.

Woodcock don't engender a lot of respect from some people. Holland (1944) refers to them as stupid in the title of his chapter about these birds, while Waterman (1972) calls them upland clowns in his book's chapter on woodcock.

In the spring, males perform an exciting mating dance in the sky each dusk and dawn. The dance is done to attract females, but once people have seen this aerial ballet, they quickly become as interested in the performance as all the hen woodcock in the area.

I owned my farm for two years before learning that the sky dance is to be seen over my woods every evening in April and May. Since we discovered it, my family and I have been reluctant to miss even a single performance. (Leopold 1949)

Today, woodcock often live in habitats heavily influenced by human activity, and their populations have declined as activities like logging, unpopular with many, have declined on the landscape. Woodcock and species that share the same habitats are frequently dependent on land use patterns.

The American woodcock is a member of the shorebird family, Scolopacidae, but it doesn't have the long legs or long neck of many of its cousins—it has very short legs and practically no neck at all. Other shorebirds travel in flocks; inhabit open marshes, wetlands, and beaches; and nest on the prairies or tundra far to the north. Woodcock got lost somewhere on the species' evolutionary journey and ended up as solitary migrators, living in the woods, and don't get much farther north than the Great Lakes and New England regions.

The closest relative of the American woodcock, *Scolopax minor*, is the European woodcock, *Scolopax rusticola*. The European woodcock is a quarter to a third larger than the American woodcock and has horizontal stripes across its chest, whereas the chest and belly of the American woodcock are a solid russet red. In the United States, the woodcock is most often confused with the Wilson's snipe, *Gallinago delicata*.

This little bird has obviously caused taxonomists some consternation over the years. Olin Sewall Pettingill (1936) records fourteen scientific names dating from 1788 to 1923. Since then, the woodcock's scientific name has changed to *Philohela minor* and most recently to *Scolopax minor*. *Minor* refers to the American woodcock being smaller than the European woodcock. *Scolopax* is simply the Greek word for "woodcock," not nearly as romantic as "bog lover," the interpretation of *Philohela*.

> The scientific binomial changed at least 14 times before it stabilized on our current [American Ornithologists' Union] checklist as *Philohela minor*. (McCabe 1982)

> When I started this book, the accepted Latin name for the American woodcock was *Philohela minor*, which translates into "little lover of the bogs." A year later the taxonomists changed the name back to *Scolopax minor*. Since 1788 the bird has been assigned no fewer than ten different names. (de la Valdène 1985)

Not everyone is tolerant of these scientific reclassifications.

> An odd bird indeed is my friend *Philohela minor*. Oh yes, I know the taxonomists recently changed his name to the uneuphonious *Scolopax minor*, but I'm an old curmudgeon now and I'm damned if I'll obey orders. Known

colloquially as timberdoodle, bogsucker, even mudbat, the woodcock is the strangest, dearest bird I know, and I won't change his name on a whim of mere science. (Jones 2002)

To add to the confusion, Pettingill writes that the woodcock has at least thirty-nine vernacular names, including (hold your breath) snipe, cock, little woodcock, less woodcock, timberdoodle, big-headed snipe, mud snipe, red-breasted snipe, blind snipe, big mud snipe, wood snipe, big snipe, brush snipe, thick-necked snipe, wall-eyed snipe, cane snipe, owl snipe, bar-capped snipe, whistling snipe, little whistler, whistler, Labrador twister, pewee, hill partridge, mountain partridge, swamp partridge, night partridge, night-flit, night peck, woodhen, mudhen, bog sucker, bog bird, bog borer, marsh plover, holumpake, shrups, night becasse, and Massachusetts woodcock. He goes on to list the names of the woodcock in Pennsylvania German, Chippewa, Canadian French, Creole, and German. Needless to say, there's a lot of confusion around this bird. Burton Spiller calls them little russet fellers, and Frank Woolner calls them whistledoodles. Timberdoodle seems to be the most commonly used nickname today.

Pettingill wrote *the* monograph on woodcock, beginning with the following description.

> The American woodcock, largely because of its retiring habits, failed to make an impression on the early colonists. (Pettingill 1936)

That's not necessarily the most flattering way to introduce a species. Woodcock did make an impression on some early colonists, but perhaps not in a favorable way.

> ... white men soon found that where woodcocks most abounded, there was malaria too.... Hence, where there were woodcocks there were swamps; where there were swamps, there were mosquitoes; where there were mosquitoes, there was malaria, although no one in that unscientific age guessed the connections. Easton, on the Delaware, was praised as a healthy place to live, "for its exemption from the fevers of the country, from

the fact of there being no woodcock ground within five miles of the Court-house." (Bakeless 1961)

This would be the statistical phenomenon of correlation but not causation. Both woodcock and the mosquitoes that carry malaria like low wet places, but neither causes a place to be wet. While we no longer blame woodcock for any diseases, they remain largely unknown and vastly underappreciated.

> To the rustic lad, and the farmer, upon whose land, among the glades and swamps, they breed, they are unseen, unknown. (Jarvis 1890)

> . . . his habitat may be within a short distance of a house, and the owners of the said house may know naught of it. (Sandys 1890)

> The woodcock is a bird of mystery to the automobilist and rural resident. People who were raised right next door to his model little nest never have seen him except when they visited the city and paid five dollars to see him without his feathers. (Askins 1931)

The reason they are so little known? They live in habitats few people would willingly walk in.

> There is no bird of which country-people are more ignorant than the wood-cock, as they are seldom seen by any except those who go in quest of them in their wet and often dreary haunts. (Lewis 1906)

> One who desires to make his acquaintance must penetrate into the depths of the most tangled swamps to find him at home. (Grinnell 1910)

> Woodcock don't come to us. We have to go to them, and where they hang out are places most wouldn't choose for a pleasant stroll. It might be an aspen thicket, young alders, cedars, hardwoods, sycamore jungles, or wil-low brakes. (McIntosh 1996)

> We learn early on the treachery of alders. The footing is inevitably mis-erable in an alder brake, the wet, mucky ground littered with the rotting

corpses of expired or moribund earlier growth just waiting to trip you up. You step on the leaning trunk of a seemingly sound alder to hoist yourself over and it breaks just as the woodcock flushes, depositing you and your gun facedown in the muck—an intimate introduction to the realities of the mud bat's world. (Jones 1996)

Dreary haunts, tangled swamps, thickets, or treachery of alders are not places for a nice afternoon stroll.

However, those who seek out woodcock are almost always rewarded, whether watching their mating dance far overhead in the spring, stumbling on a brood of dirty cottonballs with toothpick beaks in the summer, or being startled by a twittering flush in front of a dog's nose in the fall.

There is a debate in the literature about the culinary attributes of woodcock. It seems people either love them or hate them. Lundrigan (2006) refers to them as "flying liver," and from his further descriptions the reader gets the impression that he isn't a big fan of liver. On the other hand, Davis (2004) calls woodcock "the finest table bird of all, the kind of sublime fare that will bring a brave gourmet to his knees." The true gourmand, especially in Europe, lets nothing go to waste. Some prefer to eat the woodcock's bones, brain, entrails, all but its feathers.

For many, the return of woodcock is as much or more a sign of spring than the first wildflowers in the woods, first skeins of geese overhead, or first robins on the lawn. Thoughts of hearing that first peent are one of the things that get some of us through the long northern winters. Because the bird is one of the earliest spring migrants, the vigil for the first peent often finds the anxious observer shivering in the chill of a March sunset. But the thrill of the spring's first woodcock is worth several fruitless evenings just watching and listening.

Some seasons the woodcock sings before the first chorus of the spring peepers. It is the early voice of this returning migrant that marks the winter's end. (Teale 1974)

... hearing my year's first woodcock song marks the certain end of a long New England winter. (Tappley 2000)

I know the woodcock as a harbinger of spring, more cheering to me than the most mellifluous thrush. (Fergus 2005)

If this book spurs you to no other action, I hope it stimulates enough interest that you bundle up some spring evening after supper, drive to the nearest woods and walk down the road, or just step out your back door to listen enthralled while you stare skyward as a male woodcock serenades and performs for both you and any attentive ladies in the area. Bring a kid with you. Kids are always looking for excuses to stay up past bedtime. Watching the sky dance is the best excuse ever.

1 anatomy and behaviors

The American Woodcock is the oddest-looking land-bird in North America. Its legs are too short for so large a body, its tail is only half as long as it should be, its neck is too short and too thick, and its head is entirely out of drawing. The eyes are too far back, and the bill is too long and too straight. In appearance, the Woodcock looks like an avian caricature. —*Hornaday 1904*

When talk turns to woodcock, it's mostly a discussion of adaptation and evolution. —*Carney 1993*

Fact is, the woodcock looks like something designed by a committee and assembled in the dark by a mad mechanic using leftover and apparently unrelated parts.—*McIntosh 1996*

According to the Seneca Indians, after the Maker finished creating all the creatures of the Earth, He looked around and realized He had some leftover parts. There was a small pile of feathers—not big flashy ones, for those had been taken by the glamourous species He had already created, but drab earth-toned grays and browns. There was a head, but the brain was upside down, and the ears were misplaced in front of the boggled eyes, and the beak was disproportionately long. He found a chunky little body and stubby legs and sturdy but graceless wings. It was an ill-matched assortment of parts, but because the Maker hated to waste anything, He put them all together and called it a woodcock.—*Tappley 2000*

Revealing my own early literary beginnings, I'll add my own description and say that the woodcock could have walked straight off the pages of a Dr. Seuss book.

Camouflage

The first thing to notice about woodcock is that they are so hard to notice. It's not as if they are all that tiny; the adults are about eleven inches long, but they blend in perfectly with the leaf litter on the forest floor. More than once I've had the experience of knowing—or thinking I knew—exactly where a woodcock was. She's on her nest and the nest is at the base of that dogwood at the end of that log. I just checked her nest three days ago. I can stare and stare for far longer than I should have to before I finally see the bird.

> . . . the sitting bird is quite easy to approach, as she seems to know that silence is her best protection, and indeed it is, for the coloring of her plumage so blends with the surroundings that he must have sharp eyes to find the nest she so silently protects. (Jarvis 1890)

> David spotted the hen first. She sat facing away from us, flattened against the ground, her bill resting on the leaves and her black eyes looking straight ahead. I had missed her a dozen times, from no more than fifteen feet away. (de la Valdène 1985)

> If you can find her, which is virtually impossible except by sheer accident or with the assistance of a pointing dog, you can sometimes pet her without prompting a flush. I wish I could tell you how it feels to reach slowly down and stroke a mother woodcock's head while she sits there still as stone, her lovely eyes unblinking. I've done it; I just can't describe the feeling—except I imagine it's like laying your finger on the hand of god. (McIntosh 1996)

> I was told where the nest was in a general way and stood within a few feet of the bird for some moments knowing that it was close by, yet unable to detect its presence. Nothing appeared to my earnest scrutiny but a mottled expanse of dry leaves and branches as yet unadorned with foliage. Unexpectedly, my patience was rewarded as the outline of the bird in every detail flashed clearly into my vision. And it was right on the spot where I had looked so hard a moment before with unseeing eyes. (Gregory n.d.)

In the same manner, I've helped my friends Earl and Donna band chicks. With far more experience than I have, they have a much better search image. When one of their dogs is on point and there are three or four chicks just inches from its nose, they see the chicks almost instantly, but it takes me awhile before I see them in the leaf litter (figures 1 and 2).

Figure 1. This day-old chick has been banded and is ready to be released. Photo courtesy of Donna Dustin.

Figure 2. A brood of four woodcock in leaf litter. Photo courtesy of Donna Dustin.

According to the literature, I'm not the only one who has had these problems. Usually, it's her glossy obsidian eye that gives the hen away (figure 3). In the fall, everyone who has walked through the woods in the right habitat has had birds almost magically flush right in front of them.

I have once or twice almost stepped on one, so still did he crouch and so marvelous was his protective coloration. (Heilner 1941)

The browns, grays, and blues of his back and the top of his head blend so completely with the lights and shades of his habitat that he is apt to remain almost invisible even after you know his exact location. (Knight 1946)

Evolution has provided natural camouflage for most of the world's wild creatures, and never has the art been more thoroughly consummated than in the case of the woodcock. (Woolner 1974)

Figure 3. Woodcock blend almost perfectly into leaf litter on the forest floor. The woodcock in this photo is looking directly away with her eyes hidden. Photo courtesy of Donna Dustin.

These cryptic characteristics make woodcock challenging for artists. How do you highlight a bird in a painting or a drawing that isn't supposed to be seen? Jerome Jackson reports a conversation between George Miksch Sutton, one of the twentieth century's best bird artists, and a colleague while touring a museum.

> . . . they stopped in front of a painting of an American woodcock. Jay commented favorably, but Sutton disagreed . . . George noted: "You can see the woodcock." (Jackson 2007)

While many birds are well camouflaged, underneath those feathers is where the unique characteristics of the woodcock really are.

Nests and Eggs

Those familiar with the intricately woven hanging nest of the oriole, the smooth grass and mud cup the robin lays her eggs in, or the moss-lined nest of the phoebe will consider that the woodcock really doesn't have any nest at all. A woodcock nest is a few twigs and leaves scraped together on the forest floor into a somewhat cuplike shape. During construction and before the eggs are laid, few would even recognize it as a nest or even see a difference between the nest and the surrounding leaf litter.

Many eggs are not what we would call egg-shaped. Ornithologists generally categorize eggs into four basic shapes, with lots of variation within each shape. The first shape is elliptical, elongated with equally rounded ends and widest at the middle. Subelliptical eggs are rounded at the ends but a little more elongated and taper toward the ends, with the widest part more toward one end. The oval egg, what most people think of as egg-shaped, is rounder and larger at one end and tapering toward the other end. The last shape is pyriform (think pyramid), with a large blunt end tapering to a narrow tip at the other end.

Egg shape, like almost everything in nature, is a continuum. People like to create categories and label a species as being in this category instead of another. This leads to many debates over which somewhat arbitrary human category an egg should be placed in. Does this species have more pointed oval eggs or flatter pyriform eggs? It really doesn't matter. The birds don't read the books or participate in the debates.

Another factor relevant to ground-nesting birds especially is the color of their eggs. There are two basic questions to address. Why are the eggs colored and patterned? How do they get their color? The why question is an evolutionary issue. The how question is a physiological issue. Tim Birkhead's excellent 2016 book, *The Most Perfect Thing: Inside (and Outside) a Bird's Egg*, devotes two full chapters to these questions.

From the evolutionary perspective, egg color and pattern may be shaped by the need to be camouflaged and inconspicuous, the desire to avoid brood parasites, and the necessity for individual recognition. Individual recognition is probably most important in colony-nesting species where birds return from feeding to a beach or an oceanside cliff with hundreds or thousands of other nests full of eggs. Brood parasites—birds of the same or different species that lay their eggs in the nest of another bird in the hope that bird will raise their young—will have trouble dumping eggs in a nest if the pattern on the existing eggs is so complicated that the brood parasites can't replicate it well enough to hide their eggs among the other eggs. Obviously, for woodcock, camouflage is the best explanation of why coloration and patterning evolved through time.

The how is a question that involves only two pigments, both of which are related to the creation and breakdown of heme, an iron-based chemical in blood. Protoporphyrin is responsible for the brown and reddish colors. The other chemical, biliverdin, is responsible for the blue and green hues in the eggs of some species.

Eggshells are made of different layers, and different pigments can be laid down in different concentrations in each layer. That can explain the homogeneous base colors of some eggs. The more challenging question is how to explain the pattern of speckles and streaks.

Gosler et al. (2005) studied egg thickness as a function of pigmentation and found that the darker-pigmented parts of eggshells are thinner and that the pigments may act as structural reinforcement in these thin-shelled areas. Kilner (2006) cites an evolutionary arms race between birds and nest-egg predators and parasites. Birds are continually forced to create newer and different patterns that aren't recognized by the predators who are at the same time learning those patterns. This arms race leading to complexity is somewhat analogous to sexual selection leading to more outrageous ornaments or behaviors in males of some species (sexual selection will be covered in following chapters).

The pigments are applied in the uterus as the egg is moving down the reproductive tract. But it's difficult to understand why some eggs are speckled evenly, why some are speckled more at one end than the other, or why some are speckled while others are streaked (figure 4).

Figure 4. Four woodcock eggs in a nest. Their tan base color and brown speckles allow the eggs to blend into the surrounding leaf litter.

There may be multiple compounding factors that explain the colors and patterns we see. Birkhead (2016) ends his chapter with the statement, "For now the creation of complex patterns on the eggs of several bird species remains a future project."

Incubation

Newly laid eggs are denser than water. As the egg and embryo mature, some water evaporates through the eggshell, and some carbon dioxide given off by the embryo is trapped in an air pocket on the blunt end of the egg. Anyone who has shelled a hard-boiled egg is familiar with this pocket.

When placed in a glass of water, a newly laid egg will sink to the bottom. As the air pocket gets larger, the egg stays on the bottom of the glass but stands on its pointed end. A few days later, it will float in the middle of the water column as more gases accumulate in it. In another few days, it will float on the surface. As time passes and the egg gets lighter and the air pocket larger, it will float higher above the surface. As the embryo reaches maturity, it won't be perfectly balanced within the egg, and the egg will begin to float at an angle. The rate of maturation and timing for each species is different, but by floating an egg biologists can make a good estimate of its hatch date (Westerskov 1950; Mabee et al. 2006).

Pettingill (1936) measured fifty-seven woodcock eggs and found an average weight of 16.6 grams with a maximum weight of 20.4 grams, a little more than half an ounce. He then measured two clutches of eggs. Over the last twelve days of incubation, the eggs went from 14.6 to 15 grams down to 12.4 to 13.3 grams. For perspective, he found the average weight of a hen woodcock on the breeding grounds to be 198 grams or 7 ounces. If a hen generally lays four eggs per nest, the total weight of those eggs is the equivalent of 33 percent of her weight. In human terms, this would be the same as a 140-pound woman giving birth to a 47-pound baby. Or giving birth to four 12-pound babies four days in a row. Clearly, the bodies of these little birds undergo some amazing physiological stresses.

Woodcock eggs are approximately 38 by 29 millimeters or 1.5 by 1.1

inches in size (Baicich and Harrison 1997). Ruffed grouse eggs are 40 by 30 millimeters, almost the same size. However, ruffed grouse are three times larger than woodcock. Granted, ruffed grouse lay more eggs in a clutch than do woodcock, nine to twelve versus four. That's a lot of egg for such a small bird.

Woodcock hens incubate their eggs for approximately twenty to twenty-one days. Caldwell and Lindzey (1974) found that the hen leaves the nest on average four and a half times a day. She's away only roughly twenty-two minutes at a time. Nesting hens are basically on a starvation diet that entire time.

Marshall (1982b) found that Minnesota woodcock hen weights ranged around 230 grams or 8.1 ounces through March and April. In mid- to late April during the egg-laying and nesting period, their body weight dropped to around 180 to 190 grams. That's a 24 percent drop in weight over just a couple weeks. Sepik (1994) showed that female body weight in April was about 230 grams. By early June, after incubating, body weight was below 180 grams, a 22 percent drop in weight.

Hatching

Woodcock chicks emerge from the nest a little differently than most birds do. Mendall and Aldous (1943) note that successfully hatched eggs usually show a lengthwise split instead of a cracked line around the width of the egg, which is the way most species hatch. By identifying these characteristic cracks or splits, it's easy to tell if the nest successfully hatched eggs or was predated. Predated eggs look like something took a bite out of them.

The chicks of most bird species have what is called an egg tooth somewhere on the top of their beak, usually toward the tip. This sharp tooth acts as a rasp that cuts through the eggshell. The shortest way to cut a hole in an egg large enough to get out of is to go around the width of the egg.

Wetherbee and Bartlett (1962) found that newly hatched woodcock chicks did have egg teeth on both the upper and lower beaks. They further observed that woodcock have a unique way of hatching. The birds push the spinal processes—the bumps on the individual vertebrae of the spinal column—against the side of the shell and split the shell lengthwise. In other words, while other species crack their way out of the shell, woodcock rip their way out (figures 5 and 6).

Woodcock are quite successful at nesting compared to many other ground-nesting birds. Mendall and Aldous (1943) report a fertility rate among eggs of 98.4 percent. They studied 136 nests over six years and determined that 84 nests, 62 percent, successfully hatched at least one egg. In Pennsylvania, Liscinsky (1972) found a similar nest success rate of 56 percent.

Overall, woodcock have some of the highest nest success rates among ground-nesting birds. However, they also have one of the smallest clutches among game birds. If ruffed grouse have a dozen eggs per nest and woodcock only four eggs, presumably ruffed grouse could lose three times as many nests as woodcock and maintain the same population levels. The math and science are actually a bit more complicated than that, but this illustrates the point that if egg production is low, every egg counts.

First Weeks

Woodcock are precocial birds. Altricial species, like most songbirds, stay in the nest for days to weeks, being fed by the adults until they are able to fly away from the nest. But woodcock leave their nest within hours of hatching—it's too dangerous to stay in one place for too long. Mendall and Aldous (1943) report birds traveling as far as fifty to a hundred yards from the nest in the first two days of life. That's a long distance for a small bird with big feet. However, on average most broods travel less than thirty yards from the nest in the first two days.

When newly hatched, chicks weigh on average 11.4 grams, less than half an ounce. However, they grow amazingly fast. Pettingill (1936) found a clutch of newly hatched woodcock on a small island. He was able to find them repeatedly over the first couple weeks to weigh them. One chick started at 15.2 grams. Nine days later, it weighed 56.1 grams. Two days after

Figure 5. Nest of recently hatched woodcock. The egg in the lower right part of the nest best demonstrates the lengthwise split in the shell.

Figure 6. Woodcock eggs with characteristic length-wise splitting.

that, it weighed 67.1 grams. That's 11 grams or a 20 percent increase in body weight in two days. By the time the chick was two weeks old, it weighed 73.5 grams. That's a 380 percent increase in body weight in two weeks. Similarly, Dwyer et al. (1982) found that through the first two to three weeks, male chicks grew at a rate of 5.1 grams per day and females at a rate of 6.2 grams per day.

Horton and Causey (1981) found that chicks stay with the hen for about the first five weeks. For an additional two weeks, chicks in the same brood can be found together during the day but are often far apart at night. Two weeks later, the chicks didn't associate much during the day. At night, some chicks from the broods did use the same roosting fields. Often by their third week, they are almost indistinguishable in size from the adult birds. Chicks are able to make short flights at an early age.

How much food would we have to eat to almost quadruple our body weight in only a couple weeks? It's not just this growth we have to account for in their diet. While the chicks are growing, they also have to maintain their body heat, especially on those cool early-summer nights. That takes energy derived from food. Foraging for food itself burns a lot of calories. That requires energy derived from food. Digesting food requires energy. Some of that food is excreted as waste, showing that digestion isn't especially efficient. The young chicks must first take care of all these factors. Any leftover energy can then be used for growth. This highlights just how much the chicks have to be eating.

Flight

One of the defining characteristics for most birds is flight, and avian flight requires wings. Avian aeronautics relate to the natural history of the woodcock. Aspect ratio is the value calculated as the square of the wing length divided by the surface area of the wing. Higher aspect ratios are found in species with long, narrow wings and lower aspect ratios in species with short, broad wings. Birds with high aspect ratios, including seabirds such as the albatross and brown pelican, are designed for good lift and long, soaring flights. However, these species do not have the best agility or ability to make sharp turns.

Most shorebird wings are long, narrow, and pointed. The sharp tips at the end of each wing minimize the drag created by the turbulence of air moving over the tips (Henderson 2008). These wings are designed for flying long distances very fast while in the open.

Woodcock have the short, rounded, wide wings more typical of game birds such as quail and prairie-chickens. These allow them to launch explosively from the ground, a startle effect any hunter can relate to. They also allow the birds to be highly maneuverable as they dodge through all the stems and branches that characterize their early successional habitats (figure 7).

While this wing shape is good for moving within their habitats, it is a very poor design for long-distance migration. Woodcock have to migrate long distances on wings designed for short, fast bursts of speed. This is a classic example of evolution making do with what is available. There aren't wings that are both good for short, fast flights through dog-hair thickets of regenerating aspen and good for long-distance migration.

Evolution at Work

People often describe the woodcock's brain as being upside down. It's not really upside down; it's just rotated backward quite a bit. In most birds, the spinal cord exits through the bottom of the skull through an opening called the foramen magnum. In the woodcock, the brain is rotated back far enough that the brain stem actually exits out of the front, or ventral, side of the skull before immediately turning downward along the spinal column. Cobb (1959) found that the woodcock's brain is rotated back farther than any other species.

The reason the brain is rotated back is to make room in the front of the skull for the large eyes. The large eyes allow the birds to gather more of the available light at dusk and dawn when they do most of their feeding. Some scientists have speculated that having eyes more toward the top of the head

Figure 7. Wings of ruffed grouse (top), American woodcock (middle), and Wilson's snipe (bottom). Woodcock should have the sharply pointed wings of most shorebirds, like snipe, but instead they have the more rounded wings adapted to forest life, similar to ruffed grouse.

than most birds allows woodcock to look upward for predators while their beaks are down probing the soil for food.

Another unique feature of the woodcock? Its ears are located just below and in front of its eyes at the base of the bill. Presumably, this makes it easier for the bird to hear something wiggling near the tip of its beak, either in the leaf litter or under the ground.

While having eyes on top of the head and ears at the base of the bill sounds weird, it's a good example of how evolution works. Evolution rarely creates something new; it simply modifies what came before. All birds have ears and eyes. In the woodcock, the ears have moved forward in the skull and the eyes upward. It would be a challenge for evolution to create a third ear or eye, but shifting existing structures around is what evolution is very good at. Nothing new, just modifying an existing structure.

While Tappley states that woodcock are made from "an ill-matched assortment of parts," others see a beautiful design.

> An intelligent examination of a woodcock will prove him to be the most interesting example of nature's wisdom in planning to meet certain conditions. (Sandys and Van Dyke 1924)

> From his long bill to his broad, stubby tail, a woodcock is an excellent example of planned engineering. Everything about him is designed to suit the environment in which he lives. (Knight 1946)

> The bill, which at first glance appears ungainly, is a wonder of engineering. (de la Valdène 1985)

Words like "intelligent," "planned engineering," and "designed" make biologists very uncomfortable. Evolutionary biologists are actually much more comfortable with phrases like "an ill-matched assortment of parts." When engineers design something, they can design exactly what they need. They can use any piece or part from any design ever created in the past, or they can design a brand-new part that has never existed before and plug that part into their design. Evolution doesn't work like this.

Evolution can work only with what it already has—the mismatched parts—the genes in a population and the anatomical structures controlled by those genes. Evolution can take what's available and tweak it, but it can't create something from scratch. Evolution can take an existing part of the anatomy, shrink it, enlarge it, or shift it slightly, but it can't create something completely new. The individual or the population that individual lives in simply doesn't have the instructions, the genetic code, to invent something totally new.

Stephen Jay Gould devoted much of his career as an evolutionary biologist and writer to debunking creationist thinking about evolutionary design. One of his most famous essays on this topic is called "The Panda's Thumb" (1980). Opposable thumbs are supposed to be a uniquely human trait, but Gould describes the opposable thumbs he observed in pandas at the National Zoo. After watching them grasp a bamboo branch and slide their paws up it, stripping off all the leaves, he writes, "Odd arrangements and funny solutions are the proof of evolution—paths that a sensible God would never tread." "Odd arrangements and funny solutions" perfectly reflect what many have written about the little timberdoodle.

> . . . you wonder if some blunder was made in his design. He really looks like something that should be sent back to the drawing board. (Kimball 1970)

> Its parts are highly specialized and when put together turn out as a sort of caricature, with a snipe's beak, a quail's breast, and flimsy legs that no one would want. (Waterman 1972)

> With all due respect to the diminutive bird, I sometimes think woodcock are living proof that God has a sense of humor. Here is a bird with a Pinocchio-like beak, toothpick legs, and bulging eyes. It appears an upland oddball, a biological misfit left over from some mixed-up evolutionary gene pool. Yet, it is remarkably adapted to life in the forest. (Dorsey 1990)

> Gould ends his essay with the following quote from Charles Darwin:

Although an organ may not have been originally formed for some special purpose, if it now serves for this end, we are justified in saying that it is specially contrived for it. On the same principle, if a man were to make a machine for some special purpose, but were to use old wheels, springs, and pulleys, only slightly altered, the whole machine, with all its parts, might be said to be specially contrived for that purpose. Thus throughout nature almost every part of each living being has probably served, in a slightly modified condition, for diverse purposes, and has acted in the living machinery in many ancient and distinct specific forms.

These words almost perfectly reflect Tappley's comments about the Seneca believing the woodcock was made from all the leftover parts.

Woodcock aren't odd. They are feathered wonders illustrating the mechanisms of evolution. But much of their odd anatomy leads to misperceptions and misunderstandings of what the birds do and how they do it.

> . . . yet no other in the list of game birds is so little understood by mankind in general. (Jarvis 1890)

> Few men, if any, thoroughly understand the ways of this shy, shade-loving, mysterious recluse, the woodcock, for of all our game birds he is undeniably the most puzzling. (Sandys 1890)

> I gravely suspect that there has been more nonsense written about the life, food, and habits of this bird than about any other American game. . . . Had I chanced to have kept a record of all the questions concerning feathered game, probably one-half of them would have been about the woodcock, for to most men he is indeed a bird of mystery. (Sandys and Van Dyke 1924)

> One of his most delightful characteristics is this ability to make people strain their imaginations and scratch their heads in wonder. (Holland 1961)

Feeding and Diets

Obtaining food is one of the primary goals of any living organism, especially animals with high metabolisms like birds. Its long, pointed bill is one of the woodcock's most noticeable features. As with all shorebirds, the long, needle-like beak allows the bird to probe the mud for food.

The long bill itself is truly unusual. First, most of us think of bills as being hard structures. Think about a sparrow cracking seeds or a woodpecker pounding on a tree trunk. In the woodcock, the tip of the upper bill is full of nerves and somewhat flexible. This allows the bird to feel worms or grubs in the soil while it is probing.

That's not even the most interesting part about the bill. The upper bill is actually hinged halfway down its length. The bird can plunge its bill deep into the soil, feel around for grubs or worms, and then open just the tip of the bill to grab its prety and pull it out of the ground.

So how did that bill get to be so long? Evolution has the explanation. Imagine a population of short-billed birds. As in all characteristics in all populations, there will be variation. Among the population of your friends, each is a different height, has a different color of hair, and has a different skills set. No two are the same.

Within our early population of woodcock, there were some with slightly longer and slightly shorter bills. Those with longer bills were able to find more food deeper in the soil, and presumably this made them healthier. These birds had more surviving young. The young of these long-billed birds again had variation within the population: some had shorter and some had longer bills than their parents did. Over thousands of generations, the birds with longer bills ate better, reproduced more successfully, and passed those characteristics on until all woodcock had longer bills. At the same time, other evolutionary forces or constraints on the bird's physiology or other parts of its anatomy kept its bill from becoming too long. Any characteristic of any species is the continuing result of countless generations of different evolutionary, ecological, and environmental forces pushing and pulling the individuals and their traits in that population.

When it comes to using that long bill, everyone knows that woodcock eat earthworms.

> I recall the day [Aldo Leopold] said, "Tomorrow looks like a good wood-cock day," and asked would I like to go on a hunt? I most certainly would. He then requested that I bring a soils map for Adams County. . . . In my brief search, I began to wonder whether a road map might not do just as well—so why a soils map? As soon as I put the map on his desk, he began, with the eyes of a histologist, to scrutinize the multicolored plates. . . . I was to learn later, on my own, that he was looking for a soil type that was the likely home for earthworms. (McCabe 1987)

In fact, virtually everything written about woodcock includes stories about how the birds can eat their own weight in earthworms in twenty-four hours.

> The food of woodcock consists principally of large earthworms, of which it swallows as many in the course of a night as would equal its own weight. (Audubon 1840)

These statements deserve a bit of investigating.

Earthworms are native to the southern or winter habitats of woodcock. However, earthworms are considered invasive species across almost all of the Great Lakes region and New England. Frelich et al. (2006) review all the aspects of forest ecology that earthworms can affect, including the distribution of chemicals in the soil; the microbial, fungal, and invertebrate communities within the soil; plant roots; and the diversity and abundance of wildflowers and other herbaceous plants. Holdsworth et al. (2007) state that "recent research suggests that invasive earthworms are important agents of change that will affect the composition, structure, and function of northern temperate forests and are a significant conservation concern for many native plant species." The research they cite and numerous other research papers fully support these statements. Earthworms alter nutrient cycles and dramatically change the herbaceous plant communities on the forest floor. To the discerning eye, a northern forest with and without earthworms looks very different. Earthworms are nothing but bad news in the northern forests, nothing but bad news for everything except woodcock.

There are numerous references to woodcock using their long beaks not to probe down into the soil but to pick through the leaf litter.

> Their food consists of various larvae, and other aquatic worms, for which, during the evening, they are almost continually turning over the leaves with their bill, or searching in the bogs. (Wilson 1839)

> . . . we watched a couple feeding for more than an hour, and only noticed them dexterously turn over the withered leaves, and every now and then probe, with the greatest facility, the rich loamy soil with their long slender bills. (Lewis 1906)

> The food of the woodcock is of limited variety, consisting primarily of the lower forms of animal life inhabiting the soil surface, the underside of fallen leaves, and rotting sticks. (Pettingill 1936)

> During dry spells, when the worms have returned to the subsoil, the woodcock must seek other foods. It then resorts to the woods, where it turns over leaves in search of grubs, slugs, insects, and larvae. (Bent 1962)

And we can imagine that long bill being quite effective for flipping over leaves. This is another example of evolution using part of the anatomy that evolved for one task efficiently for another task.

Liscinsky (1972) reports on the nutrition of earthworms. Earthworms are roughly 83 percent water. However, the tissue of different species of earthworms ranges from 54 to 64 percent protein. Any student who ever dissected a worm in high school biology class knows that a worm is basically a tube inside a tube. The inner tube, the digestive tract, is mostly full of soil.

This is not to say that on their northern breeding grounds woodcock would never probe the soil. In addition to earthworms, there are scores to hundreds of invertebrates, larvae, pupae, and eggs in forest soil. All these would make nutritious high-protein meals.

Pettingill (1936) found seeds of at least seventeen species of plants in woodcock stomachs. In a later publication (1939), he found at least thirty-eight species of seeds. One question we don't have an answer for is whether woodcock actively seek and eat plant seeds, or whether they just happen to ingest them while eating animal matter.

Pettingill (1936) also reported eleven orders of invertebrates in the stomachs he examined and in the literature. Additionally, he found six families of beetle larvae and thirteen families of fly larvae in the stomachs he examined. He later (1939) reported that animal matter made up 95.8 percent of the total food in the stomachs. In that second report, he listed forty-five families of invertebrates identified in two separate collections of woodcock. The two collections included 70 and 124 birds.

Mendall and Aldous (1943) displayed two data tables of woodcock diets. Animals made up 89.5 and 94.2 percent of the diets. When the animal matter was broken out, the first table reported that 68 percent of the diet was earthworms but also included flies, beetles, butterflies, moths, and other insects. The second table had the same data with the addition of miscellaneous amounts of wasps and spiders.

Sheldon (1967) calculated percent volume of prey species in fifteen woodcock stomachs. He found that 39 percent of the diet was beetle larvae, 30 percent earthworms, 15 percent fly larvae, and 15 percent butterfly and moth caterpillars and pupae. He also found trace amounts of centipedes, spiders, ants, and vegetation. In total, he found at least fifteen families of invertebrates in woodcock digestive tracts during the summer.

There are reports of starving woodcock feeding on grain during inclement weather. Cottam (1934) observed woodcock eating corn at a feeding station designed for bobwhite. However, we must assume that these observations are exceptions and not the rule.

Miller and Causey (1985) studied woodcock diets in Alabama in their wintering habitat. They found a similar range of types of invertebrates eaten. However, they noted that "earthworms may be somewhat less important in the diurnal diet of woodcock on the southern wintering grounds than in northern areas." It's odd that earthworms would be less important in the south, where they have always been available. In ecological time, northern earthworms are relatively new.

What exactly do these numbers mean? Some studies may count individual organisms found in the digestive tract. Others may calculate percent volume or the calories found in each type of prey organism or the protein, lipid, or carbohydrate of each group of organisms. This can cause issues in interpreting the data because often studies measure different parts of the diets in different ways. Miller and Causey reported their dietary results by volume, dry weight, and frequency as well as where in the digestive tract they found the invertebrates.

We also have to consider the time of year and the activity of the birds when we study their diets. Spending time on the wintering grounds, migrating, and raising a brood of chicks on the breeding grounds are all very different activities that may require slightly different types and amounts of food. Miller and Causey illustrated that with their study of Alabama woodcock over the winter months.

On top of all that, imagine dissecting the stomach contents of a woodcock. Oftentimes, the researchers are just looking at ground-up pieces and parts and trying to match a leg or thorax or antenna to a prey species. Pettingill (1936) writes that "the animal life which it devours is characteristically soft-bodied. Once reacted upon by the potent digestive juices it breaks down into an unrecognizable mass of food matter with surprising rapidity."

To summarize, woodcock are opportunistic and will eat a wide variety of invertebrates and invertebrate larvae. It's a gross oversimplification to say that they eat earthworms. Although they will glut themselves on earthworms if given a chance, they do have a diverse diet.

Unusual Behaviors

Much of the historical literature on wildlife will show examples of outrageous claims about the feats of some. Woodcock are no different. Woodcock are well known for probing the soil with their bills for food. However, early naturalists had more interesting ways of describing their feeding habits.

> Some sportsmen assert that when the cocks are feeding they strike their long bills into the soil, and then, raising their bodies high on their feet, they open their wings and flutter round and round until they have sunk their bills sufficiently far into the ground to reach their prey. (Lewis 1906)

At the turn of the twentieth century, much nature writing was very sentimental and often gave human emotions and other characteristics to wildlife. Eventually, this blew up into the nature fakers debate, where serious naturalists challenged the scientific validity of many of these nature writers. One of those naturalists who waded into the fray was President Theodore Roosevelt. (What would American politics be like today if we had a president who was knowledgeable enough to routinely get into heated arguments over ornithological details?)

William Long was one of the writers who took significant liberties with his descriptions of the natural world. In his book *A Little Brother to the Bear*, he details how a woodcock with a broken leg sets a cast on that leg himself.

> At first he took soft clay in his bill from the edge of the water and seemed to be smearing it on one leg near the knee. Then he fluttered away on one foot for a short distance and seemed to be pulling tiny roots and fibers of grass, which he worked into the clay that he had already smeared on his leg. Again he took some clay and plastered it over the fibers, putting on more and more until I could plainly see the enlargement, working away with a strange, silent intentness for fully fifteen minutes, while I watched and wondered, scarcely believing my eyes. (Long 1903)

Long writes that not all woodcock are capable of such acts, but this particular bird was a genius. In his book *The Nature Fakers*, Ralph Lutts describes the reaction of scientists to this story. Needless to say, they found a few logical inconsistencies. They pointed out that to set a broken leg, this woodcock would need to understand anatomy and osteology on the medical side. In the area of material science, this bird seemed to understand that for maximum strength, he needed to mix a fibrous material with a cementlike material. Such a wise bird would indeed be a genius.

Other unusual behaviors in woodcock include hens carrying their young through the forest. D. R. Johnson reported these observations in the mid-1980s, although most citations are much older. There appear to be several different ways that hens carry their chicks around the forest.

> . . . the woodcock was carrying two young birds that were clinging to her breast. Her head was held down as if she was helping support them. (Johnson 1984)

> I have never seen the young carried in either of these ways, but have seen them transported from place to place, grasped by the long slender toes of the old bird and drawn up close to the body. (Jarvis 1890)

> We must not forget to mention one other peculiarity of these birds, and this is the somewhat extraordinary habit they have of carrying their young on their backs from place to place. (Lewis 1906)

> Her legs appeared to be half-bent, and so far as I could determine the youngster was held between them. (Sandys and Van Dyke 1924)

> Nuzzling the young gently with her beak, she lifted one and flew away through the thicket. She held her baby in her bill just as a cat holds a kitten. The fourth baby was removed in the same manner. (Rutledge 1935)

Perhaps what's most interesting is not that a hen carries her young but that she does it in so many different ways: balanced on her back, grasped in her feet, squeezed between her thighs, and clinging to her breast feathers.

Although there have been several refutations of this theory of woodcock carrying their young, Tordoff (1984) gives one of the clearest descriptions of why they don't do so. He points out that woodcock and shorebirds in general don't have toes or feet designed for grasping like a perching bird. Therefore, they have very weak muscles for grasping anything. Likewise, the musculature of the chicks' beaks would not allow them to grasp the hen, and the musculature of the hen's throat would not allow her to hold her chicks on her breast.

Tordoff continues by explaining what people are probably seeing when they think they see a chick carried below the hen. Killdeer, a shorebird relative of the woodcock, are well known for their distraction display. The female feigns injury, dragging a wing and running away from her chicks when a predator is near. The predator, attracted to this already injured animal, chases the hen, leaving the helpless chicks behind. Once she has lured the predator far enough away from her chicks, the hen flies off as healthy as can be.

Woodcock have a similar display. To distract a predator from her chicks, the hen flies off with her body in a nearly upright position. Her head and bill are turned down, her legs dangle, and her tail is depressed and fanned outward. Her flight is labored and loud. Seen from some angles, it might appear that she is carrying something between her legs. Tordoff writes that when his dogs find and point broods while he is out banding, the hen frequently flies off using this display.

Yet another unusual behavior of woodcock has been described as rocking or bobbing. Sandys (1899) reports the observations of a "wise old chap" who saw a woodcock "dancing upon the ground and tapping it with his bill to induce the worms to come up." In the next paragraph, he reports that this is done "to imitate the fall of rain-drops, which brings the worms within reach." Bent (1962) writes that the bird "is said to beat the soft ground with its feet or wings, which is supposed to suggest the effect of pattering rain and draw the worms towards the surface."

Pettingill (1936) describes what may be the same behavior in more detail, stating that each foot is placed down hesitantly and the woodcock bobs its body as if flinching from stepping on a sharp object. He hypothesizes that this was a "nervous reaction resulting from fear or suspicion." He states that he has observed this behavior by hens returning to their nests after they had been flushed but not after they had left of their own accord. He goes on to cite an earlier source that claimed woodcock used the "peculiar pressing movement of the feet" to locate worms underground. Pettingill disagrees with this interpretation and suggests that the birds, being out in the open because they couldn't find food elsewhere, were nervous.

Sheldon (1967) quotes Glasgow (1958) as stating that the foraging woodcock walked in a "rumba-like manner." Others provide even more detail.

> Peering beneath the trees, there upon a carpet of pine needles we saw a woodcock, strutting about just like a turkey-cock in miniature, with tail erect, spread like a fan, and drooping wings, nodding its head in time with the movement of its feet, as though it were listening to music we could not hear. (Jarvis 1890)

Worth (1974), in an article titled "Body-Bobbing Woodcock," hypothesizes that woodcock bob "to mimic prevailing shifting shadows." Marshall (1982a) followed that with an article titled "Does the Woodcock Bob or Rock—and Why?" He suggests that "minute movements of earthworms close to the surface (or insects in leaf litter), in response to the slight changes in pressure from the rocking bird's foot, allow the woodcock to detect them by sight or perhaps infinitesimal sound." He goes on to propose that the behavior should be termed rocking instead of bobbing.

Heinrich (2016) provides a review of the early literature, his own observations, and his own hypothesis. He observed this same behavior by a bird in an open field when the bird could clearly see him. However, when he retreated, hid, and continued to observe the bird, this behavior stopped.

Heinrich suggests that the "rocking-walk display is a response to what the bird perceives as a mild potential threat situation that is not severe enough to initiate predator-avoidance tactics to disrupt it into flight or cryptic hiding." Generally, woodcock have two responses to threats. First, they can flush, which is costly energywise. Second, they can freeze and hide. However, if they are in a food-rich area and are actively foraging, freezing takes away from time they could be looking for food and thus is also costly. If the threat is not severe, this behavior is a visual cue to the potential predator that the woodcock knows the predator is there. The predator could attack the woodcock, but since the woodcock is aware of the predator, the woodcock would probably escape any attack.

While some of these explanations seem a bit silly in retrospect, they illustrate the scientific method and process. One person makes observations and develops a hypothesis around them. It was a good enough explanation at the time that others felt it worthy of publication in a scientific journal. More people make additional observations or conduct experiments that seem to disprove the current hypothesis, and then they develop their own hypotheses to explain their data or observations. This can occur numerous times. While Heinrich's article provides a strong review of past hypotheses and existing ecological theory, someone may come along in the future with an alternative to his hypothesis. That's the way science is supposed to work.

The woodcock is a bird that can set its own leg in a cast, balance chicks on its back while flying, and dance the rumba. That's some bird!

2   sky dance

Who would suppose that this bird, indifferent as he seems in the daytime to all sentiment, could sing such love-songs in the gloaming.—*Jarvis 1890*

What other bird dances in the Spring moonlight just so that otherwise decent, sane people will stay up half the night just to watch this avian madness.—*Hill and Smith 1981*

I don't know what your ideas of heaven are, but one of mine looks a lot like a woodcock singing ground on a moony spring night.—*McIntosh 1996*

Until you have witnessed the male woodcock's sky dance, you will be deprived of one of the natural world's most spectacular events. . . . No other bird in North America can match the woodcock's aerial ballet. Few equal his earnest efforts to attract a willing female.
—*Huggler 1996*

On cool spring evenings across the Great Lakes and New England regions, male woodcock take to the sky in aerial displays to attract nearby females. They should also be attracting any beginning or experienced naturalists in the area. We often don't have to go far to find them, if we know where to look.

> . . . his habitat may be within a short distance of a house, and the owners of the said house may know naught of it. (Sandys 1890)

> This little drama is enacted nightly by at least thirty pairs of woodcocks within a mile of the outskirts of Madison [Wisconsin], i.e., within a mile of thousands of people who sigh for dramatic entertainment. (Leopold 1999)

> The woodcock is a bird of mystery to the automobilist and rural resident. People who were raised right next door to his modest little nest never have seen him except when they visited the city and paid five dollars to see him without his feathers. (Askins 1931)

Even those who are outside a lot don't know of the existence of woodcock on their own lands.

> Nearly all writers mention the fact that the woodcock is often unknown to the farmers on whose land they reside. (Huntington 1903)

> He'd lived in New England for forty-odd years, yet the timberdoodle was unknown to him. Little green men from Mars would have been no less exciting. Such creatures as timberdoodles do not exist in the minds of average citizens. (Woolner 1974)

This is most surely because the birds stay hidden during the day and only come out for a few short weeks every spring when farmers are working long hours in their fields or have already retired for the evening. But even those who hear them may not recognize them.

> The twilight flight song of the woodcock is one of the most curious and tantalizing yet interesting bird songs we have. I fancy that the persons who hear and recognize it in the April or May twilight are few and far between. (Burroughs 1901)

Reading descriptions, I found it interesting to note the range of descriptions and metaphors for the sky dance.

> Woodcock are strange birds indeed. Rarely do any two men see them in the same light. (Knight 1946)

Maybe "the same light" is the key to that sentence. Because the displays occur in the half-light of evening, in the shadows between twilight and night, our imaginations may fill in what our eyes may miss.

> The cloak of night always lends a certain mystery to the doings of nocturnal birds, and more often than not their habits justify our unusual interest in them. How many evenings have I tempted the malaria germs of the Jersey lowlands to watch the Woodcock perform his strange sky dance! (Chapman 1907)

The ear could trace his flight, but not the eye. In less than a minute the straining ear failed to catch any sound, and we knew he had reached his climax and was circling. Once we distinctly saw him whirling far above us. Then he was lost in the obscurity. (Burroughs 1908)

Indeed, as evening advances and the glow on the western horizon fades, the displays become more of an exercise in listening than seeing.

Once there, you listen. In fact, most of the time your ears must guide your eyes. (Carney 1993)

Before the nightly sky dance begins, the male ghosts unseen and unheard to his singing ground. The performance begins with the evening's first peent. The peent's harsh, nasal rasp is entirely out of place in the peace and calm of the evening forest.

But no matter how expectant I am, the sudden buzz always startles me, both by its proximity and by its harshness. It sounds like the metallic scrape of a thumbnail across the teeth of a comb. It is, technically, a song. But it's hardly melodious. It's called, inexplicably, a "peent." (Tappley 2000)

Pettingill (1936) found at least seventeen other descriptions of the sound in the literature, including pink, s-p-e-a-k, peenk or blaik, ping-k, kwan-k, sneap, zeet, speet, quack, zeeip, paap, quank, spate or skape, ze-e-ee-p, pa-anck, kwank, and paif. Several of the calls woodcock make are in the frequency range that many people have trouble hearing as they grow older. This is especially true when those ears have been exposed to a lifetime of shotgun blasts. Differences in hearing ability need to be overlaid on these various interpretations of the sound.

The woodcock is a stickler for punctuality. If three or four days in a row have the same weather, clear with no wind, by the fourth or fifth evening a person can predict the first peent almost within seconds. It's fun to take a group to the woods, ask them to synchronize their watches, and predict the first peent within twenty or thirty seconds. People always laugh disbelievingly as they walk off in pairs or trios to their assigned singing ground. When everyone reconvenes at the end of the evening, the first thing they say is that they can't believe how accurate the prediction was.

The reason for this predictability is what Aldo Leopold (1949) calls the male's vanity. The male wants everything to be just right when he begins, and he is particularly aware of the light level. Although different sources give slightly different measurements, Leopold writes that the birds demand a romantic light intensity of exactly 0.05 foot-candle.

Unfortunately, Leopold's poetic description doesn't hold up to the scientific data. Pitelka (1943) recorded the first peent on six days between March 20 and 31, finding that the light intensity varied from 1.5 to 28 foot-candles. At the same time, peenting varied between twenty-four minutes before sunset to sixteen minutes after sunset. Where Pitelka's data do agree with Leopold's predictability is in the flight song. Pitelka found that the flight always began when the light was between 0.5 and 1 foot-candle. Not exactly the number Leopold came up with, but just as predictable and pretty close.

STOP! If it is springtime and you live in an area that has woodcock habitat, don't read anymore. Spend a few evenings outside listening for yourself. Record your own impressions of the sky dance. Then read the following descriptions. Who do you think did the best job? Which is most similar to your own description? If it's the wrong time of year or you're not in the right place, read on.

> It is a voice of ecstatic song coming down from the upper air and through the mist and the darkness—the spirit of the swamp and the marsh climbing heavenward and pouring out its joy in a wild burst of lyric melody; a haunter of the muck and a prober of the mud suddenly transformed into a bird that soars and circles and warbles like a lark hidden or half hidden in the depths of the twilight sky. (Burroughs 1901)

> Then in the twilight he springs into the air with wings humming like taut wires, towering in widening circles until lost in the gloom of the upper air,

whence he comes pitching down to his mate, headlong, with the sound of a spent bullet. (Askins 1931)

He flies in low from some neighboring thicket, alights on the bare moss, and at once begins the overture: a series of queer throaty *peents* spaced about two seconds apart, and sounding much like the summer call of the nighthawk.

Suddenly the peenting ceases and the bird flutters skyward in a series of wide spirals, emitting a musical twitter. Up and up he goes, the spirals steeper and smaller, the twittering louder and louder, until the performer is only a speck in the sky. Then, without warning, he tumbles like a crippled plane, giving voice in a soft liquid warble that a March bluebird might envy. (Leopold 1949)

The descent is like a falcon's stoop, except that it zigzags and the accompanying clusters of chirp-chirp-chirp-chirp rising vocal notes have a ventriloquistic character—one moment seemingly directly overhead, then behind, followed by a split second of silence before the bird splashes down from a completely unexpected direction. (Evans 1971)

[The descent] is a ventriloquial sound, difficult to locate, coming from everywhere and nowhere. (Woolner 1974)

Higher and higher in great sweeping circles it mounts above the pasture. We follow with our eyes its retreating form, often losing it in the sky. At the height of its ascent the song begins. The sweet frail twittering sound at times seems to come from all directions, the notes to shower down around us. And while the song goes on it is joined by a quavering musical strain produced by three stiff narrow feathers at each wingtip. They vibrate in the wind as the bird plunges, veering wildly, falling through the sky like a gust-blown leaf. (Teale 1974)

In the twilight they're not easy to see, even if the anxious observer is standing relatively close. But I saw the male helicopter up from his pad. He made

a wide circle without gaining much altitude; the whole time his outer primaries were singing that liquid twitter so easily recognized in October. . . .
The male circled, gaining altitude slowly at first, but then his circles grew
smaller as he flew higher. I squinted, trying to watch him ascend for as long
as possible. Finally, he disappeared from view, but his wing song continued. After he disappeared for several seconds the wing twitter ceased. The
silence lasted only briefly. As the woodcock's descent began, he turned on
his vocal cords. The sound was not unlike his wing twitter, though it was
definitely different, even more liquid in nature, a softer song. In view again,
still high, the woodcock descended so fast it was difficult for the eye to keep
up. His warbling continued, getting louder as he came closer. (Sisley 1980)

Grey clouds darkened the sky in metronomed increments, and in a whistle
of feathers the outline of the bird appeared out of the obscurity of the field
onto the thin, pale horizon. At tree level, he accelerated perpendicularly,
trilling and dancing with heartsick abandon, spiraling ever-wider arcs until
he vanished in the clouds. Moments later I watched as he quietly glided
down. . . . At dawn and dusk every night for the next six weeks, he would
prance under the stars, sometimes, as in the case of a full moon, extending
his affairs until first light. (de la Valdène 1985)

The buzzing call was infrequent, about once every ten or twenty seconds,
and lasted for three or four minutes overall. Overture complete, the bird
launched itself on twittering wings straight upward into the empyrean.
Up, up he spiraled, twittering all the way—one hundred feet, two hundred,
three. . . . Then he dive-bombed the earth in a long zigzag glide, burbling
his flight song, a series of wild liquid chirps. (Jones 2002)

When the light fades you'll hear a nasal "peent," until at last, with a twitter
of wings, he launches into the twilight. The twittering rises and falls as
he orbits under the first stars of evening, sometimes climbing to 300 feet.
Presently the twittering gives way to a liquid warble, and he drops earth-
ward, slicing and dipping like an oak leaf. (Williams 2004)

Abruptly he flushes. With a musical whistle of wings, he angles across the field. Then he begins to spiral up into the pale evening sky. He rises higher and higher until he disappears from sight, although I can still hear his distant, muted twitter. A moment later he reappears, zigzagging and parachuting back to earth like an autumn leaf on a soft breeze. As he descends, he utters a different tone, a liquid kissing note. (Tappley 2009)

The peents grew louder and closer together, and then the woodcock flushed low for several yards before fluttering up, up into the lemony sky toward pencil points of light from stars just breaking out. I lost the bird in heaven's vault, then saw it spiral widely downward like a wall-clock pendulum. (Huggler 2017)

Rising from the singing ground, the male heads skyward in a series of expanding circles. Once aloft, he gives forth a continuous series of soft musical chirps and twitters. The sound of air whistling through the outer wing feathers can also be heard. Topping out at a height of several hundred feet above the singing ground, the male begins his descent. Still vocalizing, he quickly slips and dives back to alight again at approximately the same spot where he took off. (Kletzly 1976)

In this last quote, Kletzly states that the bird returns to approximately the same spot. There's no approximation about it. He lands in exactly the same spot.

He lands beside me, almost precisely at the spot where I first spotted him. He peents, preens, struts, then flushes again full of hot-blooded passion. (Tappley 2009)

Knight (1946) discusses how focused the displaying male is. He writes that one evening he and two friends sat on a singing ground talking quietly and smoking and did not disrupt the bird at all. Estella Leopold (2016), Aldo's youngest daughter, describes how she and her siblings could sometimes get within five to seven feet of a male woodcock. I have spent several evenings on a singing ground, sitting quietly with a Labrador retriever at my side, with the male peenting just a few feet in front of me. As long as we were still, he treated us as no more consequential than a couple bumps on a log.

The sky dance occurs over your head, a black bird darting in front of black trees against a cobalt sky. With a little practice, it's easy to trace the sound and sight of the flight. It's much more difficult to see the birds once they are on the ground where everything is black. But there is quite a performance there also. An entire display occurs on the ground that almost matches the one in the sky.

> At each utterance of the *paap* the neck was slightly lengthened, the head thrown upward and backward . . . the bill was opened wide and raised to a horizontal position, and wings were jerked out from the body. All these movements were abrupt and convulsive. . . . There was also perhaps a slight twitching of the tail, but this member was not perceptibly raised or expanded. (Brewster 1894)

> In marked contrast to the ethereal flight song, the woodcock, as part of its courtship, gives a particularly harsh, nasal sound that is most commonly rendered as a peent. . . . It is a most unusual avian sound and possesses a quality of sounds that is almost impossible to adequately describe. (Pettingill 1936)

> While calling, the male maintained a stiff stance with head pulled back, wings dropped, and tail spread and held vertically. When uttering the peent note he jerked his head backward. (Pitelka 1943)

> Often it is hidden in the dusk but on the occasions when I have been able to observe it I have noted how it lifts its wings, hunches its shoulders, and jerks its head with every "Peent!" The call seems now far away, now close at hand, according to the direction the bird is pointing. (Teale 1974)

> In addition to the harsh peent sound, males also use a quieter tuck-oo

sound right before the peent. While the peent can be heard from some distance, a person must be very close to hear the tuck-oo.

Of course, the purpose of all this time, energy, and effort is to attract a female.

> The mating flight of the male woodcock is as fine an example of exhibitionism as one could wish to see. (Knight 1946)

Sheldon (1967) gives one of the best descriptions of the mating act. First, the female must select a particular male from any of several singing grounds that may be in her immediate area. When she finally chooses and visits his ground, the male raises his wings and approaches her in a stiff, uninterrupted walk; sometimes, he makes a series of quick dashes toward her. The male mounts the female from behind, wings fluttering slowly at first but becoming more rapid before the mating act is finished.

Historically, how woodcock make the different sounds during their flights has been heavily debated. There may be a couple of confounding issues here. People hear sounds differently. It's sometimes difficult to determine what sounds are being referenced in some descriptions of the sky dance. Generally, the sound that has caused the greatest debate is a sound commonly called a twitter. The debate was whether this sound is made vocally or by the wings.

> This whistle is another of the peculiarities of the woodcock which are so puzzling. It does not come from the throat or bill, as would naturally be supposed, but from the pinions as they cleave the air. I have held a woodcock by the legs and heard this whistle as it fluttered to escape, and to satisfy myself that the noise was made by the wings, have grasped it by the neck and bill and still that whistle; but when the wings stopped beating, the whistle ceased. (Jarvis 1890)

> Twenty years ago there was much discussion as to the manner in which the startled woodcock produces the whistling sound usually heard as it springs from the ground. The ranks of sportsmen were divided into two factions,

one of which held that the whistle was vocal, while the other was firmly convinced that it was produced by the wings. Oddly enough, able ornithologists, who were also sportsmen, were divided on the question—and are probably still divided, for the matter has never been satisfactorily settled. (Grinnell 1910)

Another unknown: is the twitter vocal, or is it mechanical? My friend, Bill Feeney, once clapped a net over a peenting bird and removed his outer primary feathers; thereafter the bird peented and warbled, but twittered no more. (Leopold 1949)

Woodcock can make this twittering or whistling noise during any of their flights. The following describes the sounds that fall woodcock make when flushed by hunters.

Many years ago I expressed in print a belief that the whistling sound made by a rising woodcock is produced by the bird's wings. This conviction has since been confirmed by field experience at the lake with woodcock killed during the first half of September, and in varying conditions of moult. Such of them as still retained or had just renewed the attenuated outer primaries, almost always whistled when flushed, whereas no sound other than a dull fluttering one was ever heard from any of those not equipped. (Brewster 1925)

Today, we know that the twittering noise comes from the wings. The three outer primary feathers of both males and females are smaller than in birds of similar size, and the males' primaries are especially small and narrow (figure 8). Measuring these feathers is one way researchers determine the sex of birds. During flight, the birds hold these feathers so that wind can flow between them. When this happens, the three feathers vibrate, and it's these vibrations that cause the sound. The woodcock's close cousin, Wilson's snipe, does the same thing but only uses its outer tail feathers.

Many concepts in ecology are relatively easy to grasp. We can see predation or herbivory or the evidence of it. A look out the window at the bird

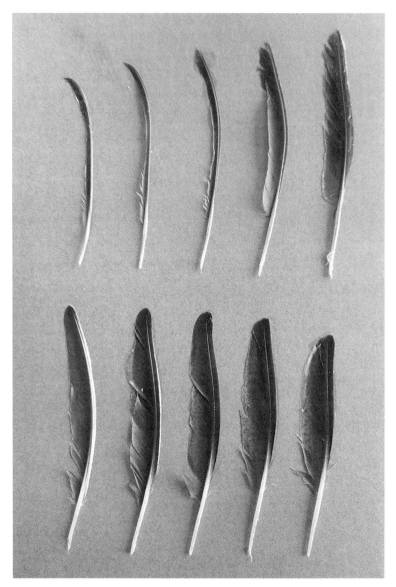

Figure 8. Five outer primary feathers (left side outermost) of American woodcock (top) and Wilson's snipe (bottom). The narrow feathers on the woodcock wing vibrate during the sky dance.

feeder is a lesson in competition and direct interaction of individuals of the same and different species. But because evolution usually occurs over such long time periods, it is often hard to perceive or demonstrate. However, in studying the behaviors of woodcock in spring, we can see or at least imagine evolutionary processes.

Female mate choice often leads to sexual selection and sexual dimorphism. A male has anatomical ornaments—bright feathers in warblers or large antlers in deer, for example—that function primarily to attract females. The ornaments tell each female that he is the best male. If she mates with him, he will pass his good genes on to her offspring.

Ornaments are a form of advertising. It takes a lot of energy to grow bigger, sing louder, or have brighter feathers than the other males. Males are telling females that they have the genes to defend the best territories and get the most food.

Ornaments may also tell females that a male has a good immune system capable of resisting disease and parasites. While other males with weaker immune systems have to spend their energy fighting parasites, he can spend that energy producing ornaments (Hamilton and Zuk 1982).

Sexual selection can lead to sexual dimorphism. Male warblers have bright feathers, females are drabber. Male deer have large antlers, females don't. The dull brown feathers in female woodcock are usually explained by the fact that hens need to be camouflaged when sitting on their nests. However, another reason female woodcock feathers are dull is because they don't have to be bright.

Female woodcock don't need bright feathers to attract a mate. They are the ones who do the choosing. Therefore, those they are choosing have to look good, be bigger, louder, or brighter than the next guy. When it comes to sex, females have the choices and do the selecting.

In woodcock, males don't show they are better than the other guy by having brighter feathers. Shorebirds in general aren't very colorful, so it would be difficult for woodcock to evolve bright colors when none of their ancestors had brightly colored feathers. Instead, they demonstrate their genetic fitness by their aerial displays. In really good woodcock habitat, it's not hard to find an area where you can hear four or more males displaying at the same time. Females can visit multiple singing grounds and males over the spring, decide who has the best display, and choose that male. Males can't choose; they can only be chosen.

> As I count, I imagine a parallel audience of woodcock females watching and listening to the males display. They almost certainly don't count as I do, but they compare the males in their own way, no doubt. To which talent or trait does she attend? (Kroodsma 2005)

Often in nature, size matters. We know that woodcock females are larger than males. But does being the largest of the males help? Perhaps not. Presumably, larger males would be more dominant and the more dominant males would have the more preferable singing grounds. Smaller, subdominant males would be on secondary singing grounds or would not have their own singing grounds at all. Do the best males get the best singing grounds? Do the earliest arriving males get the best singing grounds, and do the lesser-quality singing grounds fill up as others return over the following days? Does earliest equate to best? Are the best males the ones who do well over the winter, have plenty of fat reserves stored up, and can start migrating north the earliest? Are the males with the best singing grounds the oldest?

Keppie and Redmond (1985) harvested males off singing grounds, then harvested the males that replaced these males, and so on. They categorized each ground as being of high or low activity. The criteria for making this classification included the total number of males using the singing ground, the number of days each year the singing ground was occupied, and the amount of time over the six-year study that the ground was occupied. Presumably, the best grounds would have the greatest activity. When a male was removed, several more were waiting nearby to take his place.

They found no relationships between body weight and age class, first or succeeding males at a singing ground, or low- and high-activity grounds. They go on to state that "the lack of differences . . . is of special interest." (Often the science that gives us unexpected answers ultimately reveals more information than when everything goes as planned.) Their lack of an expected result led them to make a number of interesting speculations. Their research does a good job of demonstrating how science works. Unexpected results lead to the refinement of theories and the development of additional hypotheses. Good science should create more questions than answers.

In the forest at night, listening to several males sky dance around you, you might notice that one male peents a little louder or performs more sky dances than the other males in the area. If you go out the next night to the same area, you may gravitate to that singing ground. Now imagine a female woodcock in the area doing the same thing. The female is comparing the displays of several males and from that comparison selecting a mate. This is sexual selection occurring right before our eyes, evolution in action. If all this sounds a bit primitive, males acting silly around each other in the hopes of attracting females isn't too different from what happens at most bars.

One of the best ways to see female mate choice and sexual selection is among lekking grouse. Males come together and display side by side, and females can choose the best one. The most famous lekking species, at least in the United States, are the prairie grouse: prairie-chickens, sharp-tailed grouse, and sage grouse.

The woodcock breeding system is analogous to a dispersed lek. Each male displays individually. However, males often display close enough that females can visit several singing grounds over a night or a series of nights to select the best mate. Trying to sneak up on a male gives you an idea of what the females are sensing.

> While we are gradually gaining upon him other males are heard calling,
> and the wooded area harbors several woodcocks, each calling in turn until

the notes vibrate through the spring air like the trilling of tree toads or the singing of katydids on a mid-summer's eve. (Abbott 1914)

Whether it is the loudness or the frequency of the males' peenting or twittering, the height of their spiral, the sweetness of their song, or the duration of their nightly display, the hens have plenty of options when it comes to choosing a mate. It's ladies' choice in the spring woods.

> No one knows what makes a male a champion. Only the female knows, and it is she over eons of time who has shaped the males into the *peenters* and sky dancers that they are. By choosing a male where she can compare several at a time, she has probably even orchestrated where and when males display in relation to each other. When picking the best and most impressive males, females often agree, so just a few males in one area accomplish the majority of the matings. The genes of these supermales are then passed to the woodcock of the next generation. . . . Success begets success, and the male antics are perpetuated and accentuated. (Kroodsma 2005)

Multiple studies report that most nests are within a couple hundred yards of a singing ground. Gregg (1982) suggests a strong relationship between singing grounds, the density of nesting hens, and habitats. Are the females going to the singing grounds to nest, or are the males locating their singing grounds near good nesting habitats? The answer is probably the latter.

It's hard for us to determine what looks sexy to a hen woodcock—what attributes tell her about the ability of a certain male to be the best father to her offspring. It's just as hard for us to determine what a male looks for when he chooses his singing ground. In the end, all we really know is that some guys got it and some guys don't. We're just not exactly sure what "it" is.

All too often, there's a divide between those who watch birds and those who hunt them. Sometimes, there's even unnecessary antagonism. Birds like woodcock, as well as prairie grouse and their lekking behaviors, can attract bird-watchers and bird hunters and will be especially dear to those who are both.

> Also there's one awesome aspect of the woodcock's life that, if exploited, could bring in a lot of nonhunters into the bird's camp—the sky dance. . . . I've known brush-scarred veterans of countless forays in the fall thickets whose greater thrill is in the spring when the sun is just a garish memory and they're standing in the twilight to watch a woodcock spiral up through the steel-cold air in an ancient ritual. (Carson 1991)

> Just like that, I became a two-season connoisseur. (Carney 1993)

> You know, Tim, I believe that far more people have come to know the American woodcock through watching their spring sky dance rituals at sunset than have ever gotten to know them through hunting. (Flanigan 2013)

3 habitats

Their favorite feeding grounds are old sedge fields burned off clean, and pasture lands that have been pastured closely, but they can also be found in the cotton, corn, and cane fields.—*Grinnell 1910*

One of his favorite types of feeding areas is the sugar-cane field after it has been burnt over.—*Knight 1946*

Best of all, seek out a nice medley of alder, birch, and popple, with occasional conifers and perhaps a few junipers and clumps of blueberries. If there are some pesky briars, all the better.—*Woolner 1974*

It is difficult to comprehend what was once ideal woodcock habitat in east Texas and there are no baseline data to determine truly optimal habitat. . . . However, the American woodcock has proven to be very adept and successful in making use of current pine plantations and National Forests throughout the Pineywoods.—*Sullins and Conway 2013*

Perhaps the best word to describe woodcock habitats is "messy." "Heterogeneous" and "patchy" are also good words. Imagine walking a straight line at a steady pace through woodcock country. You should find that you're in a different habitat every few minutes. You may start in a lightly grazed pasture, continue into early successional aspen, then into alder, followed by more mature but mixed-growth forests.

Singing Grounds

Woodcock may choose a range of habitats as their singing grounds, but they are all generally in an open area, often bordered by brush. Tall trees immediately next to the singing grounds would potentially interfere with the birds' ascending flights.

Sheldon (1967) reports studies of fifty singing grounds in Massachusetts. Bluestem and meadow sweet were common herbaceous plants. However, woodcock may avoid areas with reed canarygrass, an invasive. Sheldon lists at least eight woody species commonly found on or adjacent to singing grounds. The grounds, created by people, included abandoned fields, orchards, pastures, powerline rights-of-way, dirt roads, and clear-cuts.

Mendall and Aldous (1943) report on the physical characteristics of twenty-nine Maine singing grounds. Most had a slight degree of ground cover, most commonly grass, weeds, or small bushes. The topography was flat. Nesting cover for females and diurnal cover for the male was usually less than a hundred yards away.

In Pennsylvania, Gutzwiller and Wakeley (1982) identify three habitat characteristics that best describe singing grounds. First was the height of trees along the edge. Shorter is better. Second was shrub density within the clearing. Males tended to prefer areas with higher shrub density, which might be related to predation. There were still plenty of open areas on the ground for their displays, but the higher shrub density might make it harder for hawks or owls to swoop down on them. Last was the size of the open areas. Gutzwiller and Wakeley found the average size of the openings to be 6,400 square meters, but they ranged from 43 to 56,000 square meters.

There are several ways to measure singing grounds. Pettingill (1936) writes that they can be as small as fifty square feet. That may be the area actually used while the male is strutting and peenting on the ground. We can also measure the entire area covered by the aerial part of the display.

While the birds may return regularly to the same area, that area is constantly changing through land management practices and ecological succession. Sheldon (1967) found that of sixty-seven singing grounds identified in 1951, only 24 percent were still active nine years later. Likewise, Sepik and Dwyer (1982) stated that sprout growth would render most clear-cuts managed for singing grounds unsuitable for use within three to four years of cutting. The culprit was plant succession. These numbers show how much we need to manage and continue to manage landscapes if we want woodcock to persist in them.

Spring

Woodcock hens nest in a wide variety of habitats. However, they most commonly nest near edges, relatively close to singing grounds, in brushy habitats with a high density of woody stems. They often locate their nests near the base of a tree or a clump of shrubs.

Although aspen is probably the most talked-about tree when it comes to woodcock habitats, in Pennsylvania, Liscinsky (1972) found that it was the fifth most common tree near nesting sites, ranking behind hawthorn, crab apple, elm, and alder. This makes sense as Pennsylvania is near the southern edge of the aspen's range. Liscinsky did find that the average height of trees in nesting cover was only twelve feet.

Likewise Pitelka (1943), studying woodcock outside Chicago, reports red oak, white oak, elm, ash, hickory, honey locust, sycamore, and willow in his study area. The only species he mentions that is more typically described for woodcock habitats is hawthorn.

Moving north into more typical woodcock habitats in Maine, McAuley et al. (1996) found nests in fifteen different forest types. Fifty-eight percent were in aspen, 11 percent in tamarack, and 10 percent in mature second-growth gray birch. Clear-cuts were associated with 66 percent of nests. McAuley et al. compared their data to Mendall and Aldous (1943) and found them similar. They go on to list at least twenty tree species that can be associated with nest sites and another eight shrub species. However, the two studies used different methods to classify the forest types, so it was difficult to make a direct comparison.

> The cover that immediately surrounds the nests is so varied in both type and extent as to discourage correlations. (Mendall and Aldous 1943)

Gregg (1974) studied nests in Wisconsin and found at least thirty woody species around nest sites.

One common thread through most nesting studies is that nests are near forest openings and often relatively close to the singing ground of a male, presumably the male that the hen mated with. It may also be the case that it's the *structure* of the vegetation, brushy and tangled, more than the tree or shrub *species* that is most important to woodcock.

There are a number of reports of nesting in southern states, including Alabama (Roboski and Causey 1981), Arkansas (Long and Locher 2013), Kentucky (Harris et al. 2009), and Texas (Davis 1961; Whiting et al. 2005). In Alabama, nests were much more common in mixed pine-hardwood forests than in hardwood stands or pine stands. Most nests were in open-grown pole timber and sawtimber stands. On a positive note, Roboski and Causey found that about 65 percent of the forests in the state at that time were structurally similar to where they located woodcock. The Kentucky researchers concluded that the structure of nesting habitats was similar to that reported in other studies with an emphasis on midstory and horizontal plant cover.

Summer

Woodcock brood habitats are similar to nesting habitats, with perhaps less brush. Brood habitats change quickly. Dwyer et al. (1982) found that broods less than and more than five days old used slightly different habitats in Maine. Older broods used open, more mature stands with a lower density of trees. Throughout the summer, Rabe and Prince (1982) found more than 90 percent of all woodcock contacts were within fifty meters of a clearing. Put another way, most birds were found near edges.

In Pennsylvania, Liscinsky (1965) recaptured ten banded woodcock within the same year. While three birds moved from one to five miles between captures, the rest of the birds moved less than 300 yards in the five to six months between captures. Doherty et al. (2010) found that more than 90 percent of woodcock moved less than 400 meters, and just under 50 percent moved under 50 meters. Over the first twenty-five days, the home range of broods in Minnesota was from 15.1 to 16.5 acres (Wenstrom 1974).

In Michigan, nests were found in areas with a lower tree density, near openings, and on dry, well-drained soils. Brood cover contained twice the tree density, was farther from openings, and on damper soils (Bourgeois 1977). Another Michigan study found that mixed deciduous forests were used for nesting and brood rearing, aspen throughout the summer, and immature aspen more in the fall (Rabe 1977).

Wishart and Bider (1976) studied woodcock habitats in southwestern Quebec throughout the growing season from late April to early November. From late April to early June, woodcock were most common in mixed woods. From June through early September, birds were found most often in deciduous forests. Starting in early September, birds gravitated toward alder woods.

Woodcock can change habitats even over short intervals of time. In Missouri, broods three days old used openings. By day eight they were using edges, and by day thirteen they were primarily using forests (Murphy and Thompson 1993).

Wenstrom (1974) found that woodcock broods avoided edges and areas with high herbaceous density. They were commonly found on soils that were classified as excessively drained and droughty. This seems to contradict the anecdote McCabe (1987) told about Leopold looking for hunting areas that had "a soil type that was the likely home for earthworms."

The following descriptions are more typical of fall habitats when the birds are hunted. But they all illustrate that woodcock habitats include more than just aspen.

> Even the so-called "aspen" or "alder" coverts contain a variety of other plants. Pure alder or aspen stands are not so attractive to woodcock as coverts where alder and aspen are mixed. The most productive cover I ever hunted grew on a hill in Maine and was formed by ideal juxtaposition of young alder, apple, aspen, gray birch, and white birch, sprinkled with young pines and broken by small openings. (Sheldon 1967)

> An ideal woodcock range consists of a springy alder or dogwood swamp, adjoining spotty thickets of hazelbrush, blackberry, young popple, and young oaks. (Leopold 1999)

Open areas are also a common factor in woodcock habitats during both summer and winter. Woodcock use these fields primarily for night roosting. There has been quite a bit of discussion in the literature over why woodcock spend the evenings in open areas and how prevalent this behavior is. Sheldon (1961) was one of the first to describe this behavior. In Massachusetts, he found that birds actively fed on various insects in these fields but did not feed on earthworms to any great extent.

In Maine, Krohn (1971) looked at the stomach weight of woodcock harvested at different times during the night in open fields. He found stomachs were much heavier as the birds were flying into the fields. From this, he concluded that the birds did little feeding in the fields. Masse et al. (2014) took this one step further and looked at foraging benefits and risk of predation in forests and in clearings. They found that earthworms were three to four times more abundant in daytime forest cover compared to fields and that mammalian predators were more common in the forested habitats. They concluded that woodcock spend the evenings in roost fields more to avoid predation than to eat.

Not all birds use young habitats all the time. New research is showing that birds use some older forests later in the summer. Perhaps this is related to soil moisture and worms. If other areas become drier, these darkest, shadiest mature forests may be the best place with the moistest soils for probing in.

> They could not be found as usual in the heavy timber under the largest trees, where the ground was soft and damp and cool. . . . Under the great oaks, in deep and impenetrable shade and gloom, behind fallen trunks where the sun never shone, and death was disrobing the giants of the past, we often found them fluttering up. (Merritt 1904)

Winter

With many migratory bird species, we know a lot more about their lives in the summer than in the winter. Many birds migrate south of the U.S.

border. Doing research in foreign countries is often difficult and expensive. Reproductive rates during the summer months generally drive most population dynamics. So even for species that overwinter in the United States, research during the winter months has always lagged behind research during the summer breeding periods. Fortunately, that has been changing in recent years.

Woodcock spend the winter months across the southeastern states. However, a large percentage eventually end up in Louisiana. Mendall and Aldous (1943) show the wintering range starting in southeastern Virginia; looping just to the east, south, and west of Tennessee; up to southern Missouri; over to the northwestern corner of Arkansas; and then south into eastern Texas.

However, the birds are not evenly distributed across the region. Knight (1946) writes that 50 percent of wintering woodcock are found in Louisiana, 15 percent in Arkansas, 10 percent each in Texas and Mississippi, 5 percent each in Florida and Alabama, and 5 percent in the other states. Modern research has also shown a concentration in Louisiana and immediately surrounding areas.

Some theorize that birds move only as far south as they need to. If a cold front comes through, they will move a little farther south. This is called a partial migration. Krementz et al. (1995) showed evidence to the contrary. Their data indicated that once birds arrived at a wintering site, they remained there until the spring migration several months later.

When you walk in northern habitats, it doesn't hurt to wear leather gloves and chaps. In the south, a suit of chain mail might be more appropriate in some areas. While habitats up north can be dangerous, southern habitats can be almost deadly.

The critical difference between northern and southern habitats is that the latter demand that you wear seventeen-inch-high rubber boots, briar-proof chaps, and noninsulated gloves. At home [Michigan] I sometimes look at the few scratches and cuts on the backs of my hands—those stitch marks of experience—with a certain feeling of pride. In Louisiana, if you have failed to protect yourself, you look at your lacerated and punctured body and feel pain. (Huggler 1996)

The green briars don't make small scratches. They rip large wounds. When I came out, I was bleeding from a cut on my cheek and blood was running down into my beard. I had several places on both hands that were bleeding. (Mathewson 2000)

Were I to long for hawthorns and other spiky plants, there were local varieties of brier and thorn trees. One would put the hawthorn to shame. It had thorns growing from thorns . . . these thorn clusters resemble medieval weapons. (Masotti 2015)

What isn't trying to stab you is trying to trip you.

Areas with high flush rates were usually "brushy" and frequently contained a moderate to dense growth of briars (*Rubus* spp.), greenbriers (*Smilax* spp.), trumpet vine (*Campsis radicans*), peppervine (*Ampelopsis arborea*), grape (*Vitis* spp.), or poison ivy (*Toxicodendron radicans*). (Roberts et al.1984)

Southern habitats have one more threat that makes them dangerous, more for people and dogs than woodcock: venomous snakes. While those who study or hunt woodcock in the north don't need to worry too much about this factor, the swampy areas frequently described for southern woodcock are also good places for several venomous snakes. During the winter hunting months, snakes are often lethargic or hibernating, but it's still something to think about.

Mendall and Aldous (1943) describe at least three primary types of forested habitats for woodcock in Louisiana. These include boggy thickets, old-growth bottomland hardwood forests, and swamplands. Thickets and swamps don't sound like much fun, but old-growth forests sound like nice places for a walk. Mendall and Aldous go on to say that the old-growth forests must have extremely thick ground cover dominated by greenbrier. Tree species common in different woodcock habitats include redgum, overcup oak, bitter pecan, cypress, tupelo, longleaf pine, slash pine, loblolly pine, and shortleaf pine.

McHenry (1983) describes several types of forest habitats for Louisiana. In the southwestern part of the state, woodcock primarily use bottomland forests. In the northern part, they use mature upland pine forests divided by streams. Clear-cuts make up a third habitat type. He goes on to write that there seems to be a relationship between woodcock and creeks bordered by beech trees.

McHenry found the largest concentrations of woodcock in young pine stands three to seven years old. More importantly, it's not just pine. He goes on to list eleven other shrubs and trees common in these cutover habitats. The final characteristics of these habitats are "impenetrable areas" of greenbrier, blackberry, and oak saplings.

Roberts et al. (1984) studied woodcock in the Mississippi delta and found that most birds were in regenerating forest stands, but a significant number were also in mature bottomland hardwood forests. Merritt (1904) frequently mentions hunting woodcock along the Mississippi River or on islands within the river.

Berry et al. (2010) classified southern woodcock habitats as having sparse ground cover and good overhead cover. They found three forest types were preferred: early successional forests, thinned pine sawtimber plantations that were regularly burned, and mixed pine-hardwood sawtimber stands. More importantly, they noticed a shift in habitats over their two year study between upland forests and floodplain forests. One year of their study was wetter than the other, and the birds moved to where the soil conditions were right.

Similarly, while some habitats are better or worse for woodcock, Sullins and Conway (2013) state that during the winter months, the birds can be found in most forest types and age classes in eastern Texas. In other words, woodcock are pretty good generalists during the winter.

This brings up another aspect of habitat management. We need to manage for different seasonal uses, but we also need to take into account wet and dry years and make sure there is some habitat in the area that woodcock can use with different rainfall and temperature patterns.

As with any wildlife species, woodcock are often just going where their prey are. While woodcock do prefer earthworms, they have a varied diet in winter just as they do on their breeding grounds. In Louisiana, Dyer and Hamilton (1974) found that woodcock diets included beetles, fire ants, millipedes, and earthworms. In Texas, Gregory and Whiting (2000) identified beetles, beetle larvae, centipedes, grasshoppers, and spiders in woodcock diets.

Forests aren't the only habitat types woodcock use in the southern states. Pastures are also commonly used. However, the birds will avoid tall rank grass.

> Pasture land is used for feeding to a large extent but pastures must be just so in order to attract woodcock. The birds show a remarkable preference for grass that is about four or five inches high. Long grass does not prevent enough freedom of movement and cuts down range of vision. Short grass fails to provide enough cover. (Knight 1946)

Additionally, burned grassland gives woodcock the open habitat structure they prefer.

> It appears that fire plays an important role by making an otherwise unsuitable habitat attractive to woodcock. (Johnson and Causey 1982)

Working in Georgia, Berdeen and Krementz (1998) found that 49 percent of nocturnal observations were in fields. Fields needed both bare ground for ease of foraging and some vegetation at the three- to six-foot level, which presumably provided some protection from predators. One- to three-year-old clear-cuts seemed to be favored by woodcock. Horton and Causey (1979) found similar numbers in Alabama.

Krementz et al. (1995) also found a similar percentage of use of fields in Virginia. In both Georgia and Virginia, they noted that some fields were heavily used by woodcock and some fields were barely used at all. This shows the importance of landscape. Fields have to be the right size and have the right vegetation and management, and they have to be in the right location on the landscape if woodcock are going to use them. When managing woodcock habitat, we have to take the large landscape view.

Especially in the winter and along the migration routes, there are numerous references to woodcock using the edges of cornfields. Jarvis (1890), Sandys (1890), Huntington (1903), Lewis (1906), Chapman (1907), Grinnell (1910), Sandys and Van Dyke (1924), Bent (1962), Vale (1936), Betten (1940), and Sheldon (1967) all talk about the birds using cornfields.

A number of researchers have recently studied agricultural fields. In Arkansas, Krementz et al. (2014) found the highest densities of woodcock in unharvested soybeans. Blackman et al. (2012) noted that in North Carolina, woodcock densities were highest in no-till soybean fields planted after corn and undisked cornfields that had a ridge and furrow topography. The furrows provided both concealing cover and more food.

In their next study, Blackman et al. (2013) found that 94 percent of nighttime locations were in forests and 6 percent were in undisked cornfields or no-till soybean fields. They hypothesize that some of their forest locations could have been openings within the forests.

This use of row crop fields reflects many early accounts of hunters finding woodcock in cornfields. However, the cornfields of the early 1900s were very different from the fields of the early 2000s. Modern fields are repeatedly treated with chemical herbicides and pesticides, many of which have a surprisingly long half-life (the time it takes for 50 percent of a chemical to break down). Today these habitats may offer little in the way of worms or invertebrates to forage on, and any earthworms out there may have high levels of chemicals. While one worm's load of poison isn't enough to cause problems, eating hundreds of worms night after night may lead to significant cumulative ingestion of a range of chemicals.

Just as in the northern states, habitats in the southern states require repeated management to remain optimal for woodcock. Johnson and Causey (1982) found that woodcock primarily use longleaf pine habitat only in the first two years after a prescribed fire. Fire cleans out the undergrowth and burns through the leaf and needle litter, exposing more of the ground and allowing the birds better opportunities to probe. Roberts et al. (1984) encourage managers to thin tree plantations as soon as possible to encourage the growth of mid- and understory vegetation.

Western Birds

Woodcock are forest birds, and therefore their western extent is where forest and prairie meet. Unfortunately, not all the birds read the same books.

Sheldon's (1967) western limit of the woodcock starts in northwestern Minnesota, angles down to the eastern half of Iowa, then goes almost straight south, including an eastern sliver of Texas. Olmstead (1951) records woodcock in Lawrence, Kansas, in early March. Judd (1937) compiles several records of woodcock in North Dakota, while Aitken (1938) records November woodcock in Fayette and Clayton Counties in Iowa. Smith and Barclay (1978) use these and other observations to redraw the western extent starting about a third of the way east into North Dakota, straight down through central Oklahoma, and then slightly east to roughly the Houston area.

One issue with mapping distributions is that people generally don't look for woodcock on the prairies—prairiedoodles, as it were. They aren't supposed to be there, so no one looks for them. It's sometimes surprising what you find when you look somewhere you're not supposed to, according to the books. For all species, the next question is, What are their densities on the edge of their range?

Are there enough individuals to make a sizable contribution to the population? Or are there just a few individuals hanging out on the fringe of the species' range? At some point when biologists are drawing lines on a map, they have to determine where a reasonable population density is and draw the line there. Any species will always have a few individuals on the other side of that line. If there are enough, then it might be worth redrawing the line. If biologists tried to track down every individual of a particular species and make sure the line includes them, they would never get anything else done.

While I have never searched for woodcock on the prairies of Minnesota, I have found several nests and randomly flushed other birds. The nests I've found have generally been on wet brush prairies, usually at the base of a shrub, often dogwood. However, I did find one nest in the middle of a very dense stand of big bluestem. It looked like classic pheasant habitat.

> Still, birds en route in small numbers are often flushed in open woods, and sometimes discovered right in the marshes of the central prairie regions of Illinois in what should be by all rights sacred jacksnipe territory.
> (Ripley 1926)

Throughout the day, season, and year, woodcock need a wide range of habitats. Just as important, those habitats need to be relatively close to each other so the birds can get from one to the other quickly and easily.

4   historic hunting and modern threats

There is but little pleasure to be obtained from summer cock shooting. It is very hot, tiresome work at best, hard alike on man and dog.—*Huntington 1903*

In these six days we averaged over sixty woodcock per day between us.—*Merritt 1904*

Summer cock-shooting is a very weak imitation of genuine sport. The birds are in poor condition—moulting flutterers, merely able to weave a batlike flight through a tangle of sun-parched foliage.—*Sandys and Van Dyke 1924*

Woodcock generally travel at comparatively low levels. Numbers of dead birds picked up beneath telephone wires and at the base of buildings and lighthouses after a night of migratory flights are mute evidence of this fact.—*Pettingill 1936*

Many conservationists have heard the early stories of hundreds of ducks or prairie grouse being harvested in a short time. However, in the nineteenth century, shorebirds were also harvested at a high rate. In fact, one of the species of North American shorebirds driven to extinction was the heavily hunted Eskimo curlew. The last birds probably didn't die in front of a gun, but the massive amount of hunting surely played a role in their extinction.

Unlike woodcock, most shorebirds travel in large flocks, often made up of multiple species. Shorebird hunters could set out a few stick-up decoys on a beach during the spring and fall migration and harvest a large number of birds. With a single shot, often a dozen or more birds could be hit. Duck hunters know that after the first volleys, ducks usually leave the immediate area quickly.

However, when crippled shorebirds thrash around on the sand, their actions seem to attract the rest of the flock, who often circle back and hover above them. Hunters could continue to shoot at these birds. Using this method, birds could be harvested in almost unimaginable numbers. Levinson and Headley (1991) state that in the late 1800s "wheelbarrows full of curlews, plover, and whatever the winds brought in that day were emptied into barrels and taken by the railroad to market." In his classic *The Wind Birds*, Peter Matthiessen (1967) writes, "Under the circumstances, one wonders that any shorebird survived into the present century."

There was a strong demand for shorebirds. Leopold (1949) describes the "lure of plover-on-toast for post-Victorian banquets." The tastiness of these birds is related to their migratory physiology. Flying that far requires a lot of energy stored as fat. Crowell (1947) notes that "we called [Eskimo curlews] Doughbirds, as they were so fat sometimes when shot in the air they burst open when hitting the ground." Such a bird would baste deliciously in its own juices.

Most of these hunts were done on nice flat sandy beaches where the birds came to the hunter and the habitat made for easy walking to pick them up. Woodcock were also harvested in relatively large volumes in the "good old days." However, because they were more solitary and occupied less friendly habitats, the volume probably didn't reach that of some other shorebirds.

Hunting Woodcock

In the north, woodcock season used to start in early July when birds barely old enough to fly were shot. In fact, some females could have still been sitting on nests or had young that were only a few days old. Remarkably, there was quite an outcry in the sporting literature against this practice. This was in the day when spring hunting of migratory waterfowl was still accepted and legal.

> As for summer shooting, it is cruel and wrong, both in theory and practice, and no manner of logic can make it right. (Jarvis 1890)

> No sportsman can take any particular credit for himself for the wholesale slaughter of young cocks during the month of July. (Lewis 1906)

Gunners turned out on the first day of the hottest months and floundered through the swales where, bathed in perspiration, stung by insects, and cursing the suffering, sluggish dogs that lolled and staggered in the moist heat, they shot young woodcock fluttering up before the guns on wings so newly feathered as to scarcely support the awkward, tender young bodies. (Sheldon 1993)

Frank Forester was one of the most famous of nineteenth-century hunters. He recounts how, when he was with a friend one summer day, the friend shot a woodcock. A few minutes later in the same spot, his dog located a "young downy, unfledged woodcock, less than two inches long." Further down the page, he reveals his reaction to the shooting of the hen. His comments at first could be regarded as noble.

Had I needed anything to convince me that woodcock ought not to be shot in July, that scene would have convinced me; and since that day I have never ceased to advocate a change and simplification of our game laws which should prohibit the killing of woodcock until the first day of October. (Forester 1951)

However, a page later, Forester reveals himself, and probably other hunters, to be a bit of a hypocrite when it came to summer shooting.

For the present, however, until the game laws shall be altered, and established on a more reasonable and more permanent footing, of which I flatter myself there is still a remote hope left to the true sportsman, there is nothing left to do but to make the best of it, to take to the field ourselves and do our best at the slaughter.

In the winter months, birds were hunted with night lights—torches were carried in front of the hunter. Hunters shot at eyeshine and harvested hundreds of birds per night.

The shooter, armed with a double barreled gun, and decked in a broad-brimmed palmetto hat, sallies forth on a foggy night to the "ridge," where

the cocks are now feeding in wonderful numbers. His companion on these expeditions is generally a stout-built negro, bearing before him a species of old-fashioned warming pan, in which is deposited a goodly supply of pine knots. Having arrived on the grounds, the cocks are soon heard whizzing about on every side; the pine knots are quickly kindled into flame, and carried over the head of the negro. The shooter keeps as much as possible to the shade, with his broad-brimmed palmetto protecting his eyes from the glare, and follows close after the torch-bearer, who walks slowly ahead. The cocks are soon seen to be sitting about on the ground, staring wildly around in mute astonishment, not knowing what to do, and easily knocked down with a slight pop of the gun . . . with an experienced "fire-hunter" it is no unusual occurrence to bag in this way fifty couple before morning.
(Lewis 1906)

Woodcock and Wilson's snipe are the only shorebirds that are still hunted. However, today that hunting is regulated at both the state and the federal level. The harvest is small compared to what it must have been back in the market hunting days. In the 2016–17 season, 26,600 hunters harvested 44,400 woodcock in the Eastern Management Region. In the Central Management Region, 104,000 hunters harvested 158,000 woodcock. (I discuss these regions in chapter 6.)

In 2016–17, the Eastern Management Region was led by Maine, New Hampshire, and Vermont with a harvest of 6,700, 6,600, and 5,300 woodcock respectively (figure 9). Among the states in the Central Management Region, Michigan, Wisconsin, and Minnesota had the highest harvests at 64,900, 35,100, and 25,900 birds (Seamans and Rau 2017).

Modern Threats

Many of the threats faced by woodcock today are the same ones faced by other migratory species. In recent years, the collision of birds with windows, powerlines, wind turbines, and so on has received a lot of popular attention. From the dates of the following quotes, this is a topic people have been aware of for at least a century.

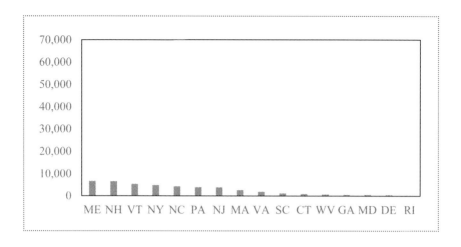

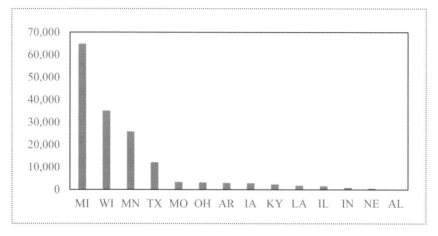

Figure 9. The number of woodcock harvested in states in the Eastern (top) and Central (bottom) Management Regions in 2016.

This flight is low and swift, and the birds are often killed by flying against telegraph wires, and sometimes dashing themselves against buildings. (Grinnell 1910)

The vast network of telephone, telegraph and trolley wires that is now stretched over the country is perhaps the greater menace to the Woodcock than any other bird. Many years ago Audubon observed that the Woodcock migrated at night, and flew very low . . . many hundreds of birds are picked up under the wires. (Forbush 1916)

From 1960 to 1974, at least six articles were published on the effects of DDT on woodcock, primarily in eastern Canada. Pesticides can have a number of different and sometimes compounding effects on wildlife. First, animals may be directly sprayed. Both insecticides, which are often toxic to the nervous system, and herbicides, some of which have been linked to a number of health issues in humans and presumably wildlife, can cause serious problems. Second, because insects are the primary prey of many birds, if a large percentage of insects are killed, there is very little for birds to eat. Third, the insects the birds do find may themselves have been exposed to sublethal levels of pesticide. By the time a bird eats thousands or tens of thousands of insects over a few months, those chemicals can build up to amazingly high levels in its system. This could be especially troublesome for hens about to lay eggs. Fourth, many people remember the stories of eggshell thinning in top predators such as bald eagles, peregrine falcons, and ospreys in the late 1960s from DDT. Chemicals can often have unintended and unexpected consequences for wildlife.

Finally, many insecticides are neurotoxins. Unfortunately, the nerve cells of insects, wildlife, and humans are remarkably similar. Anything that affects the nervous system of an insect can potentially affect the nervous system of nontarget species. This can be especially problematic for very young birds, especially those who haven't even hatched. The nervous system does a lot of development while the young chick is still an embryo in the egg. Normal development will be difficult if the hen ingested neurotoxic insecticides and transmits those chemicals to her eggs. Oftentimes, wildlife face all these issues together.

Wright (1960) found that among woodcock in DDT-sprayed areas, only 23 percent of the harvested birds were young-of-the-year birds. In unsprayed areas, 45 percent were young-of-the-year birds. In other words, it appeared that DDT spraying cut woodcock reproduction in half. Wright also found decreased reproduction in the year following spraying, suggesting that the effects linger.

In a follow-up paper, Wright (1965) reported that 86 percent of woodcock sampled contained either DDT or heptachlor in their tissues. Heptachlor was sprayed primarily in their southern or winter range, while DDT was primarily used in their summer habitats in the north. Further, Wright found that birds carried significant amounts of both chemicals with them when they arrived on the breeding grounds in the early spring. These chemicals would have been carried in their bodies from the previous summer or winter. More importantly, they were passing these chemicals on to their eggs— the chicks were exposed before they even hatched. Heptachlor in human food has a zero-tolerance level, so these birds aren't even safe for human consumption if harvested by hunters.

It gets worse. Dilworth et al. (1974) reported DDT levels of 25.4 parts per million in woodcock. Seven ppm is the threshold set by U.S. and Canadian authorities for humans. The situation was so bad that the woodcock season was closed for a couple years in some areas in the early 1970s because DDT levels in their meat presented "an unacceptable potential hazard to human health."

Working in the birds' southern range, Stickel et al. (1965) fed heptachlor-contaminated earthworms to woodcock. Six of twelve birds died within thirty-five days, and another four birds died by the fifty-third day of the experiment. On the breeding grounds, Dilworth et al. (1972) found a small but significant decrease in eggshell thickness in birds exposed to DDT. However, they concluded that thinner eggshells alone were not enough to explain the decline in productivity found in other studies.

We all know that DDT was banned. Actually, it wasn't. American chemical companies still produce large quantities of it. They just ship it to other countries. So we don't have DDT in the United States anymore. Problem solved, correct? No, DDT is just one of hundreds of pesticides used in agriculture, lawn care, and so on.

In recent years, a new class of pesticide has been developed from the nicotine molecule found in tobacco. These chemicals are collectively called neonicotinoids, neonics for short. Initial tests indicated that they might be better for the environment and for people. Nicotine is, after all, a naturally occurring plant compound. Additionally, neonics are often applied directly to crop seeds instead of sprayed in the environment. Together, that should make these chemicals more environmentally friendly. However, further tests and experience showed neonics to be more harmful than initially thought.

Goulson (2013) provided a review of the current literature on these insecticides. Researchers have found that often seeds absorb only 2 to 20 percent of the chemicals. The remaining 80 to 98 percent leach into the soil. From the soil, they can leach into ground and surface waters. Once in the soil, the chemicals' half-life can be as long as 1,250 days.

Douglas et al. (2015) show how these chemicals can move through a food chain from prey to predators within the soil community. While woodcock may not be heavily exposed to these chemicals on their northern breeding grounds, they do fly over and often winter in areas with fairly intensive agriculture. If the birds get a larger percentage of their diet from invertebrates in the soil, the chemicals will kill many of those invertebrates, and the ones that are left will probably pass on a dosage to woodcock.

Another problem is lead. Hunters of big game, such as deer, fire a single bullet from their rifles. It's a single piece of metal. Bird hunters use shotguns. Shotgun shells are loaded with hundreds of small pellets that spread out when they leave the barrel of the gun. The bird then flies through this cloud of pellets. Because it usually takes only a couple pellets to kill most birds, the rest of the pellets land somewhere past them. And even very good hunters hit only a percentage of the birds they shoot at. In the case of a miss, all the pellets end up on or in the soil.

Lead shot was banned from waterfowl hunting in the early 1990s, but it is still legal for upland game in many areas. The biggest problem with lead is having it in the environment. After decades of waterfowl hunting, soil under many wetlands had significant amounts of lead shot. Waterfowl would accidentally ingest some of these pellets. Lead itself is relatively inert. However, if a lead pellet is ground up in the digestive tract and then exposed to digestive acids, it can quickly enter the bloodstream and cause a host of problems with the nervous system and other parts of the body. There are no safe lead levels.

Waterfowl hunters tend to concentrate their activities around wetlands, which also concentrates the lead shot. It was always thought that upland hunters were spread out enough that the shot would be widely dispersed and wouldn't cause problems. That may not be the case.

Scheuhammer et al. (1999) found elevated levels of lead in woodcock in eastern Canada. A second investigation (Scheuhammer et al. 2003) tried to determine where that lead was coming from. They identified three potential sources: residue from mining, lead that persists in the soils and environment from before the ban on leaded gasoline, and ingestion of lead shot.

In their second study, they collected young-of-the-year woodcock from several locations in Ontario, Quebec, New Brunswick, and Nova Scotia. They looked at the isotopes of lead found in the birds. Although they did not go as far as specifically blaming lead shot for the elevated levels of lead, they did say that their data were "not consistent with exposure to environmental Pb [lead] from past gasoline combustion nor Precambrian mining wastes but was consistent with ingestion of spent Pb shotgun pellets."

We can draw some preliminary conclusions from this study that even when lead is scattered across large upland acres, there is enough for wildlife to accumulate it. Another aspect of this isn't talked about enough. Everyone who has eaten wild game has occasionally bitten down on a pellet. Presumably, some of those pellets miss our teeth and go right on down the esophagus. There the lead will mix with stomach acids, and from there it's straight into the bloodstream. Lead shot can be as much a human health issue as it is a wildlife health issue.

Managing Young Forests for Woodcock and Other Species

One of the biggest threats to woodcock is maturing forests, as we will see in the later chapters. And woodcock aren't the only species of wildlife with this problem.

"Best Management Practices for Woodcock and Associated Bird Species" (Wildlife Management Institute 2009) includes twenty-five species listed as likely to respond to early successional forest management and another twelve listed as having variable responses to this management. These are actually limited lists as they include only species classified as Species in Greatest Conservation Need by Michigan, Minnesota, or Wisconsin. Thirteen of these thirty-seven are listed as such in all three states.

> "I had carried into my professional life a desire never to see a tree cut, but having seen the bird diversity jump by a third following the harvest— especially some troubled species like field sparrows, prairie warblers, and woodcock—that emotional wrestling is gone. You walk through those transitional zones in spring now and hear a great deal more chatter than you used to. Now all I think is, 'Where can we do more?'" (Tim Mooney, Rhode Island Nature Conservancy, quoted in Freeman 2013)

> "We were kind of surprised at the number of species that were using, in fact had their highest abundance in, clear-cuts or the scrubby growth that comes in 5 to 10 years after a clear-cut," says Hagan. "The birds included species of concern to conservationists, such as chestnut-sided warbler, Nashville warbler and Lincoln's sparrow." (John Hagan, Manomet Center for Conservation Sciences, quoted in Lipske 1997)

Reading the literature on other wildlife, we see phrases that could have come right out of the woodcock literature.

. . . recent redstart population declines have resulted from losses of early successional habitat. (Hunt 1998)

Natural forest succession . . . may explain the current decline in [rufous-sided] towhees. (Hagan 1993)

With little effort, thousands of acres of golden-winged warbler habitat could be created in New England and New York simply by allowing lumber companies to clear-cut small areas in state and national forests and leaving the land to regenerate on its own. But it's an unpopular idea. (McLeish 2007)

Scientists often categorize bird species as young—early successional or young forest—or mature forest species. Both these terms are obviously relative and don't have a standard definition that applies to all forests and all wildlife. That categorization is usually based on the nesting habitat of that species. It might be better to say early successional nesting and mature forest nesting species. However, real life is messier than our simple classifications.

Several other eastern birds depend on brushy thickets. This habitat is not usually the subject of Sierra Club calendar photos or poetry, but these birds, and a multitude of other species that depend on low shrubland and vine tangles, are an important part of the biological diversity of North America. (Askins 2001)

. . . mature-forest birds can be viewed as a major component of early-successional bird communities, a pattern not generally recognized by ecologists. (Vitz and Rodewald 2006)

Many species or groups of species don't use the same habitats for all parts of their life cycle. For migrating species, winter and summer habitats are often different. Migratory habitats may look different from summer and winter habitats. In a surprising number of cases, the fledging habitats of juvenile birds in their first summer and the habitats adult birds use while molting in mid- to late summer are quite different from the breeding habitats the adult birds were using just a few weeks earlier in the summer.

There could be several reasons for this. The thick vegetation characteristic of early successional habitat provides a lot of hiding and escape cover when it comes to predators. Juvenile birds still learning to fly and molting adults are possibly a little slower or less adept at maneuvering. A sharp-shinned or Cooper's hawk can easily fly through a mature forest but would have more trouble navigating through the dense vegetation of early successional habitats.

Early successional habitats often have lots of fruit-bearing vines, brush, and trees. Raspberries and blackberries will quickly sprout after an area has been clear-cut. Within a few years, chokecherries, pin cherries, plums, dogwoods, and other fruit-bearing shrubs will be common (Vitz and Rodewald 2007). The seeds of some of these species can sit in the seed bank for years or decades waiting for a disturbance. They can then grow rapidly, taking advantage of the sunlight. Clear-cuts and early successional forests are often sunnier than mature forests and therefore warmer. These warmer areas can attract more insects (Smith and Hatch 2008). Additionally, insects will be attracted to the flowering fruit shrubs early in the season and the fruit itself later in the season.

Because all these factors occur at the same time, it's often difficult to tease apart which one or ones are the most important to wildlife species and what they are responding to most strongly. More importantly, different wildlife species respond to different features in clear-cuts. Porneluzi et al. (2014) and Schlossberg and King (2009) studied the response to clear-cutting over time of both early successional bird species and mature forest bird species. Each species responded differently to the vegetation in the different ages of clear-cuts. Some species peaked in abundance in the first couple years after the clear-cut, while other species didn't peak until the clear-cut was several years old.

One songbird closely tied to woodcock habitats is the golden-winged warbler, another species that has been doing very poorly in recent years. In fact, some woodcock banders will listen for golden-wings in the spring to determine if it's worth having their dogs run through an area. If they hear golden-wings, they know there's a good chance of finding woodcock.

However, we face the same issues when managing for golden-wings as we do with woodcock.

> [Golden-winged warbler researcher John Confer] supported a plan sub-mitted to the New York Department of Environmental Conservation to do some clear-cutting in the Hudson River Valley to create shrub habitat, "but it blew up in their faces and they had to retract it," he said. "Politically, you just can't clear-cut." . . . It seems that some of the land uses hated most by the general public are turning out to be the best land uses for golden-winged warblers. (McLeish 2007)

It's not just birds that respond well to early successional forests. Writing about elk management and potential reintroduction in parts of Minnesota, Albert (2017) says, "Both Michigan and Wisconsin successfully reintro-duced elk. Biologists there told [Fond du Lac Band wildlife biologist Mike] Schrage one key to success is young aspen."

Moose are another member of the charismatic megafaunas that require some young forests. In a sixty-three-page review of moose habitats in Min-nesota, Peek et al. (1976) come to a number of conclusions. They begin the "Management Implications" section with "it seems redundant to state that the increases in this moose population appear to correlate with logging activities, since this has so frequently been observed across North America and Eurasia."

If we leave the young aspen of the northern forests behind, we see this same general pattern with other species. In the northern New England states, the Canada lynx is a federally listed species. The best way to manage for lynx is to manage for their prey, snowshoe hares. Where there are hares in abundance, there will be lynx. Hares like clear-cuts. Sharp (2015) quotes Mark McCollough with the U.S. Fish and Wildlife Service as saying that one of the goals with hare-lynx habitat management is to create vegetation "so thick you can't walk through it." Lynx and hares are dependent on coniferous trees such as spruce and fir instead of the aspen, birch, and alder that woodcock like, but the same principles apply. They need very dense vegetation, the kind that is shaded out and dies as the forest matures.

Lynx and their habitats face the same issues with public perceptions that woodcock and their habitats do. Silvestro (2007) notes that "the kind of management that local forest enthusiasts are likely to endorse and the kind that help lynx may be widely divergent things." Nearly the same comments occur in the woodcock literature.

The Kirtland's warbler, which has spent a number of years on the Endangered Species List, was recently delisted. Kirtland's warblers are dependent on jack pines for their nesting, and jack pines are dependent on fire. Rapai (2012) writes that "today humans see a stand of charred tree trunks and tend to think of it as an eyesore or a tragedy. But these natural disturbances are a fundamental part of the ecology of the Kirtland warbler's ecosystem. The warblers and the other plants and animals in this area have not just adapted to fire; they have come to depend on it." A few pages later, he notes that "Kirtland's warblers nest on 'a working landscape'—an area that requires extensive and ongoing intervention, cultivation, and renewal."

Moving south into the habitats that some wintering woodcock may use, McFarlane (1992) says that the red-cockaded woodpecker, another species that has spent time on the Endangered Species List, is "truly a victim of benign neglect. . . . The landscape grinds inexorably toward a climax hardwood forest. The woodpecker is a mid-successional forest species which has lost its most valuable ally: wildfire."

All these species need early or mid-successional forests. They and their habitats all face the same issues with public perceptions when it comes to active management of working lands. In fact, it's remarkable how similar the comments are about needing to constantly manage—that is, disturb— habitats across types of species and forests. Comments about the interface of social pressures and negative perceptions with wildlife and early successional habitat management are also remarkably similar.

Where does this leave us relative to woodcock? Masse et al. (2015) wanted to know if woodcock habitat management benefited other species. They surveyed bird diversity at known singing grounds and at random locations. The total number and diversity of birds varied but were always at least 1.5 times greater on singing grounds. In this way, woodcock may serve as an umbrella or flagship species for many other species. Umbrella or flagship species have numerous definitions. The basic argument is that instead of trying to develop fifty management plans for fifty species, we find the one species that most encompasses all the habitat requirements of those species, we manage for that one species, and by default we manage for all fifty. It helps if that one umbrella species is charismatic and generates a lot of public support.

This is not to argue that all forests should be cut! We need mature forests. We need large tracts that never hear the roar of a chain saw or see a road. Many species of wildlife depend on these habitats. And many people like to hike or camp in these areas or just like to know they are out there.

Another way to look at this is to look at three "wood" birds: woodcock, wood ducks, and pileated woodpeckers. Woodpeckers, especially pileateds, like dead or diseased trees that are infested with insects or their larvae. Those are often large old trees. Wood ducks need trees mature enough to have cavities large enough for them to nest in. Often, those cavities were created by woodpeckers. Then we have woodcock, who really like young forests. All three of these species are obviously important. We can't manage every acre for all three. But we need to make sure that somewhere on the regional landscape, there are significant amounts of all the forest types capable of supporting good populations of each species. Obviously, this principle can be expanded to all wildlife species.

There are also different types of forests. Clearing tropical rain forests or harvesting old-growth forests in the Pacific Northwest is fundamentally different from managing aspen forests in the Great Lakes or New England region. Cut these forests, and it may be thousands of years before they return to their precut condition and can host the same wildlife species. As we've seen with aspen, they can recover in a matter of months to years, not centuries to millennia.

We can argue that not all tree harvesting is bad. When done carefully and thoughtfully, it can be very beneficial to a number of wildlife species. Some species are dependent on it. As a side benefit, it can be equally as beneficial to local economies.

What species do we want our management activities to encourage? Do those species need just a little more habitat management or a major habitat program across an entire region? How will we mitigate any damage to the other wildlife that use the habitats we're managing? How will this management affect the local or regional economy? How will the neighbors view it? None of these are easy questions to answer. Come to think of it, that's becoming a recurring theme to the chapters of this book!

5 studying woodcock

No one knows anything about woodcock. He's the least understood of any game bird.
—*Holland 1944*

To scientists, woodcock represent a taxing challenge; to me, the very fact that they puzzle the best experts is but another of the bird's charms. Enigmas are revitalizing.
—*de la Valdène 1985*

There are few other game birds which occasion such affection in those who study them.
—*Vance 1981*

Studying woodcock is like studying foxfire.—*McIntosh 1996*

Wildlife biologists are often in out-of-the-way places at odd hours doing things that may appear strange to some. Consequently, almost all wildlife biologists have at least one run-in with the law at some point in their career. More than once, I've been stopped and asked what I'm doing. I'm surveying woodcock, or prairie-chickens, or songbirds. This most often leads to further questions. "What's a woodcock?" "I didn't know they were around here." "Why do you want to do that?" It's always good to have a pair of binoculars, a clipboard, and a data sheet handy to prove that you really are a scientist counting birds. Usually, after a brief and polite conversation, the officers shrug their shoulders, wish you luck, and move on to look for other troublemakers.

Developing Surveys

In a perfect world, from a biologist's and a statistician's perspective, wildlife would be easy to see and count, would be equally visible in all habitats, and wouldn't move around much. Almost no species matches that description, and woodcock are especially challenging.

> If ever a bird was cloaked in mystery, that bird is the woodcock. Nocturnal
> in his habits, he is not only shy and secretive during the shooting season,
> but manages to keep his movements concealed throughout the year.
> (Spiller 1972)

There is no perfect wildlife survey. We can't count every individual. We can only do that when a species is endangered and only a handful of individuals are left—not a desirable situation. Surveys don't even try to count every individual out there. Imagine counting every duck on every wetland across the Prairie Pothole Region of the Dakotas and Prairie Provinces. Imagine counting every songbird in the eastern forests. It can't be done. What scientists are trying to get is an index. An index of a population tells them that one area has more birds than another area, numbers have increased or decreased from last year, or this year's index is above or below the five-year, ten-year, or long-term average.

Surveys usually require that data over large areas and preferably long time frames are aggregated. One survey along a few miles of road in one corner of one county done one time every spring tells us almost nothing. Data from hundreds or thousands of surveys compiled across large regions and many years can become a powerful research and management tool.

A lot of science goes into developing surveys. First, we need to determine the different types of habitats across a region. Then, we need to make sure the surveys will adequately sample all these different habitat types; we need to put surveys in good, average, or poor habitat for the species being surveyed. For instance, if we put surveys only in the very best woodcock habitat, we would get pretty high numbers because there would be a lot of birds there. If we then extrapolated these high numbers to the entire region, we'd think that woodcock were incredibly abundant.

> Currently, softwood forest appears to be sampled insufficiently to track
> woodcock population trends across all usable habitats in New Brunswick.
> (Keppie et al. 1984)

Thus, one would speculate that on the basis of woodcock habitat, the survey routes are not randomly placed and probably reflect higher numbers of singing grounds than are present in randomly located areas.
(Shissler and Samuel 1985)

Even if woodcock are detected infrequently in average or poor habitat, a large amount of that habitat on the landscape may account for a large number of birds or a large percentage of the birds in that region. For woodcock that need transient habitat, a survey site that has always been poor in the past may become very good if a clear-cut or prescribed fire creates young forest habitat. Alternately, as the vegetation at a good site matures over the years, woodcock activity will decline.

Researchers must also consider the biology of the species when designing surveys. If the surveys are done too early in the season, they may detect a number of migrating birds moving through the area and not be a true index of the number of resident or breeding birds. If the surveys are done too late, breeding activity will have declined and the birds will be much harder to detect.

The detectability of a species is the next issue. Many bird surveys are done as routes. The surveyor stops along the route at specific intervals and listens or watches for a specific period of time. If the points are too close together, there is a chance that a bird could be double-counted at two sequential points. If they are too far apart, the survey could miss birds along the route. Ideally, surveyors could spend lengthy amounts of time at each point on the route. However, there is usually a set amount of time to run the entire route, so each point can be sampled only for a limited time. Researchers have carefully designed woodcock survey routes to account for these factors.

With woodcock, the average duration of flight or the sky dance is about fifty-six seconds (Duke 1966). Therefore, two minutes at a point should be enough time to hear any peenting between flights. In the same study, woodcock peenting could be detected at a range of 82 to 257 yards, depending on the listener and the conditions. Today, survey points are four-tenths of a mile or 700 yards apart. Therefore, each point is surveying all the birds in a circle with a radius of 350 yards around the point.

The time of year the surveys can be conducted are broken into twenty-day blocks in five narrow zones running west to east across the northeastern quarter of the United States. From southern Illinois to Virginia, surveys should be done between April 10 and 30. From northern Minnesota across the Great Lakes to northern Maine, the survey period runs from May 1 to 20.

Duke also measured the stable period of detectability each night at about forty-two minutes. Therefore, the entire route needs to be run in roughly that time frame. The instructions for modern surveys allow for two minutes at each stop and roughly a minute to drive to the next stop. With ten points on the route and three minutes to survey each point, including travel time, the entire survey lasts thirty minutes. On clear evenings, surveys start twenty-two minutes after sunset.

Researchers also need to determine the sources of variation they can expect to deal with. Is the survey done before or after leaf-out? The rustling of aspen leaves can be distracting. What is the wind speed? Are there background noises such as traffic, frogs calling, or farm equipment? What is the hearing ability of the surveyor? What is the skill level of the surveyor? Statisticians constantly struggle with the best way to deal with these variables.

Singing-Ground Surveys

It would be nice if every singing ground had a woodcock and that woodcock stayed on his singing ground and only his singing ground all season long. Unfortunately, that isn't the case. Davis (1970) recorded two males on adjacent singing grounds. One woodcock ran off the other woodcock, returned after five minutes, and for the rest of the night alternated his sky dance from the two singing grounds. He switched between singing grounds for the next two evenings. After those nights, he stayed on his original singing ground, and the second ground remained vacant.

Shissler and Samuel (1983) recorded a similar situation where a color-banded male chased off a neighboring male and used two singing grounds. Later in the spring, a second male established a territory on one of the singing grounds, and the original male stayed on his first singing ground.

This movement isn't rare. When Sheldon (1967) retrapped forty-eight males a week or more after their initial capture, twenty-three or 48 percent were recaptured on a different singing ground. These recaptures were up to three and a half miles from the first capture. At six sites, two males were captured on the same singing ground on the same night.

However, the same study did show that birds will return from their long migration to the same general area. Of seven adult males originally banded on singing grounds, six returned to singing grounds within a mile and a half of the ground they were first banded on, and the seventh was recaptured ten miles from his initial capture.

The U.S. Fish and Wildlife Service's Singing-ground Surveys are an effective method for tracking woodcock populations (figure 10). Shissler and Samuel (1985) found a 98 percent correlation between birds identified during a survey and known birds along the survey routes. Nelson and Andersen (2013) noted that the cover types most often associated with woodcock were found along Singing-ground Survey routes in approximately the same abundance of those types on the landscape. They concluded that the survey is a reliable tool for long-term monitoring of the relative abundance and population trends of woodcock across the Great Lakes region. There is a correlation between singing males and broods using the same habitat (Rabe and Prince 1982), again implying that these surveys do represent overall population levels.

Once we know the methods are reliable, scientists can take these data and use them to model woodcock distribution and continue to refine our estimates of local, regional, and national populations. Thogmartin et al. (2007) used Singing-ground Survey results combined with land cover data to identify the relative abundance of woodcock across the northeastern United States. They were able to predict several major concentrations of woodcock, including east central Minnesota; the intersection of Vermont,

New York, and Ontario; the Upper Peninsula of Michigan; and St. Lawrence County in New York. They go on to say that their results "provide a basis for the development of management programs and model and map may serve to focus management and monitoring on areas and habitat features important to American woodcock."

The Latest Science

Scientists don't just keep doing these surveys and hoping for the best. They are constantly analyzing data and analyzing how they analyze data. This is serious stuff, and no one takes it lightly. A lot of resources are spent conducting these surveys, and the results of the surveys govern some pretty important decisions.

One of the best ways to evaluate older methods is with newer technology. One new technology that ornithologists studying many species of birds have started to use is stable isotopes. Isotopes are atoms that contain dif-

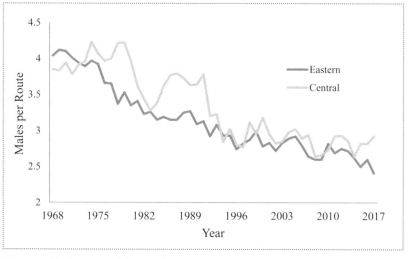

Figure 10. Singing-ground Surveys have shown a steady decline in woodcock from 1968 through 2017. Adapted from Seamans and Rau 2017.

ferent numbers of neutrons in their nuclei. For instance, carbon by definition will also have six protons in each nucleus. However, a carbon nucleus could also have six, seven, or eight neutrons. Carbon 12 has six protons and six neutrons, while carbon 14 has six protons and eight neutrons.

This becomes interesting to ornithologists because different parts of the world have slightly different ratios of different elements. Because birds molt in predictable patterns at specific times of the year or specific ages, researchers can remove certain feathers from a bird in one part of the world, study the isotopic signature of that feather, and determine where the bird was when that feather grew in.

Sullins et al. (2016) studied the first primary and third secondary feathers of birds of known origin. From these feathers they were able to calibrate an isoscape, a landscape based on woodcock feather isotopes. They then used hatch-year feathers to determine the origin of birds harvested from across the northeastern United States. They found that only 48 percent of juvenile woodcock originate in areas currently covered by the Singing-ground Survey. Fifteen percent of these birds may have bred to the north of the survey area.

If researchers decide that they need to modify a survey—adding new areas, for instance—these types of research projects are invaluable in guiding their decisions.

The Wingbee

Each fall, hunters harvest woodcock from across the eastern United States. Some of these hunters remove a wing from the birds they harvest, place it in an envelope that the U.S. Fish and Wildlife Service's Wing Collection Survey previously sent them, record basic information like date and location, and send the envelope to the Harvest Survey Branch of the Fish and Wildlife Service's Patuxent Wildlife Research Center in Maryland. Researchers there record the data from the outside of the envelope and give each envelope a bar code. The wings come from long-term participants in the program as well as from hunters selected through the Fish and Wildlife Service's Harvest Information Program.

The envelopes are then boxed up and sent to the location of the wingbee, which is held in late winter or early spring each year. In even-numbered years, it is hosted by a southern state. In odd-numbered years, a northern state is the host. Biologists from federal and state wildlife agencies and conservation organizations get together for a week to determine the age and sex of the thousands of wings that the hunters sent in. They descend on the location on a Monday. The evening is spent catching up with old friends and meeting new ones.

The real work begins the next morning, when they get to the room where they'll do the work. More than forty boxes packed with envelopes greet them. Before beginning, everyone takes a short quiz to refresh themselves on how to determine the age and sex of a woodcock from its wing. First-timers get a crash course before taking the test.

Sheldon et al. (1958) did some of the first work on aging woodcock by their wings. Martin (1964) showed how looking at the secondary feathers on the wing can determine age. For determining sex, Mendall and Aldous (1943) and Artmann and Schroeder (1976) measured the length of the wing chord. Greeley (1953) demonstrated how the widths of the three outer primary feathers can also be used to determine sex.

In 1994, Greg Sepik put all this information together into a twelve-page booklet titled "A Woodcock in the Hand." This booklet will be the bible for the duration of the wingbee, as it tells participants pretty much everything they need to know for the rest of the week. The booklet also fits nicely into hunters' game vests, so they can immediately determine the age and sex of birds they harvest while still in the field.

First, they determine the sex of the bird each wing came from. Female woodcock and their wings are significantly larger than males. Biologists determine the size of almost all bird wings by measuring the wing chord. On the wrist or bend on the front of the wing, there is a small notch that you can see or feel with your fingertips. The distance between that notch and the tip of the longest feather on the back side of the wing is the wing chord.

If the wing chord is less than 125 millimeters long, the bird is a male with

100 percent accuracy. Wing chords between 125 and 127 millimeters are 95 percent accurate for being male. Wing chords 139 to 145 millimeters or longer are 95 to 100 percent accurate for being female. Wing chords between 128 and 138 millimeters are uncertain. Hunters who have the bird in the hand can measure the length of the bill. Less than 66 millimeters is a male, and greater than 69 millimeters is a female. However, wingbee participants have only the wing.

The other way to determine sex is to measure the three outer primary feathers. These are the feathers that vibrate and twitter during the sky dance or when the bird flushes. Measurements are taken 2 centimeters from the tip of the feathers. If the combined width of the three feathers is 12.4 millimeters or less, the bird is a male. If the feathers are 12.6 millimeters wide or wider, the bird is a female.

Next, they need to determine the age of the bird. For most birds, we can tell if they hatched this year—if they are hatch-year or juvenile birds—or if they are at least one year old or older—after-hatch-year or adult birds. After they are a year old, we can't determine their age. For instance, we can't determine whether a bird is two versus three years old.

There are two different ways to identify the long feathers on the backside of the wing, often called the flight feathers. The outer ten feathers are called the primaries and are numbered from the middle of the wing to the tip; the outermost feather is P10—the tenth primary—and we number inward to P1. The long feathers on the back of the wing closer to the body are called the secondaries. These feathers are measured toward the body. The feather in the middle of the wing is S1—the first secondary—and S9 is closest to the body. Sometimes S7 to S9 are called tertials instead of secondaries.

However, to simplify things, many people just number from the outside of the wing toward the bird's body. In this case, feather 1 is P10 and feather 11 is S1. The wingbee uses the simplified version and focuses on feathers 15 to 18.

Juvenile feathers have a light band at the tip and a distinct dark band just below the light band. The mottling on the feather is symmetrical on both sides of the rachis, the rib that goes down the center of the feather. Adult feathers have little to no mottling, and any mottling is also symmetrical on both sides of the rachis. There is no distinct dark band near the tip of the feather.

It all sounds pretty easy. Sometimes it is. Sometimes it feels like you can figure everything out in the time it takes the wing to fall from the shaken envelope onto the table. For other wings, there's a lot of head scratching and consultation with those nearby. By the middle of the third day and well into the process, the floor of the room is usually covered with stray feathers. So is everything else in the room. But it can't be all work.

Get any group of biologists together, and there's going to be a field trip. Field trips are the best chance for the meeting hosts to show off their recent habitat projects and exchange ideas with colleagues. They are great opportunities for all the visitors to see different habitats or management approaches. There's a lot to learn from the formal presentations on the trip and even more to learn from the casual conversations. It's always fun to see how different parts of the world are both similar to and different from your own part. This is another good reason to move the wingbee around the country and from north to south. Biologists get to see a lot of the country and a lot of different habitats if they attend more than a couple years.

Another highlight of the trip is one of the evening meals, where the hosts show off their local cuisine. Louisiana hosted a crayfish boil. Texas had a barbecue.

After they've cleaned up, packed up, and gone home, biologists get to look at the data. Some of the most important data to come from the wingbee concern the age ratio. A high percentage of adults indicates a poor year for reproduction. Conversely, a high percentage of juveniles indicates a good year.

For the 2016–17 hunting season, 1,110 hunters submitted wings from thirty-six states. The top states were Michigan (239 hunters), Wisconsin (174), Maine (111), and Minnesota (99). Hunters submitted 4,577 wings from the Eastern Management Region and 6,753 from the Central Management Region (Seamans and Rau 2017).

One of the frustrations of scientists is that many people believe that research takes the fun out of nature. Scientists just reduce everything to numbers. Yes, they do. But those numbers reveal so much. In his 2005 book *The Singing Life of Birds: The Art and Science of Listening to Birds*, Donald Kroodsma sits and watches woodcock for several nights, turning their sky dance into a series of data points. And yet you can read in every line how much fun he's having.

While others may simply say, "That's neat," Kroodsma goes on to show what those numbers reveal about this one bird on this one night as well as the bigger ecological and evolutionary picture of the entire species. He's out there because he loves what he's doing. And note that the subtitle of his book includes both the words "art" and "science."

Banding Birds

One May, I met Earl Johnson, retired Minnesota Department of Natural Resources manager from the nearby Detroit Lakes office, and his setters midmorning at Tamarac National Wildlife Refuge in northwestern Minnesota. Several of us spent many April evenings marking peenting grounds in this part of the refuge. Today we start the next important (and fun!) work: banding broods of woodcock. If we're lucky, we'll band a few hens also.

We'll be following in the footsteps of Andy Ammann, who largely pioneered this technique and published a booklet titled "A Guide to Capturing and Banding American Woodcock Using Pointing Dogs" in 1981. The dogs seem more anxious to get going than we are. Most hunting dogs spend the spring taking sedate morning walks, then lying around the house or kennel for the rest of the day. Today these dogs have work to do. It's the same work they do every fall, work they were bred for and love. They have to find a woodcock and freeze on point. From the hunter's perspective, there may be no better type of off-season training than woodcock banding.

We'll rotate a couple dogs today, but we start with Earl's most experienced dog, who leads us into the woods. Earl follows the dog, and I follow Earl, dragging with me a rather large long-handled net that was not designed for wiggling through young aspen whips and scrub vegetation. I spend more time turning around to unhook the net from a branch than I do walking forward. The rest of the time I spend retrieving my hat, which is repeatedly knocked to the ground or hung up on a branch.

The area we're searching today looks like a photograph from a textbook on nesting habitat. It's also where we watched several peenting males through April in the adjacent openings. In a remarkably short time, the dog locks on point, Earl quietly sneaks up behind the dog, and I creep up behind Earl. We have a brief whispered discussion, Earl pointing to the hen and her four chicks. It takes my less practiced eyes a little longer to find them. While we can easily scoop the chicks up by hand, it would be nice to catch the hen also. I slowly move the net into position over her. As I quickly try to slam the net down on top of her, of course it catches on more twigs. As she sails out from under the net, I picture her laughing at me.

At least we have the chicks. This is one of the most exciting and stressful times in the life of any biologist, when they are holding the animals they are studying. Our goal is to work as quickly, quietly, and carefully as possible to minimize stress to both chicks and hen, who is undoubtedly not far away watching us. Earl places his hat over the chicks. We get out bands and a small ruler. First, we place an aluminum leg band on the chick and write the number down in Earl's notebook. We then measure the length of the bill.

Woodcock bills are approximately 14 millimeters long at hatching, and we know that the bill grows at a rate of 2 millimeters per day. Banders measure the bill, subtract 14, and divide by 2 to determine the chick's age. For instance, this chick has a bill 18 millimeters in length. Subtract 14 from 18 to get 4. Divide 4 by 2 to get two days old. This method works for the first fifteen days. After that, bill growth slows, banders need to refer to charts to estimate the age of the chick, and the estimates are less precise (Ammann 1982). We weigh the bird, put a numbered aluminum band on one leg, and repeat the steps three more times. As Earl processes the birds, I record the information. We place the four chicks together, gather our equipment and the dog, and hurry off. We hope the hen will fly back to her chicks immediately. When we are a suitable distance away, the dog starts searching again.

Banding woodcock chicks seems odd. Surely when those baby legs grow, the bands will constrict them? To add to their comical anatomy, newly hatched chicks, although tiny even compared to their small mother, have legs and feet almost the size of adult birds. One wonders how they can walk without tripping over themselves. However, this does allow biologists to put bands on very young birds without risking injury as they grow.

Between 1960 and 2016, biologists in the United States and Canada banded more than 126,600 woodcock (figure 11). The heyday for woodcock banding was the late sixties and early seventies. The peak year was 1972, with more than 6,300 birds banded. Woodcock have been banded in forty-four states and provinces. Michigan, where Andy Ammann lived and worked, leads the way with 35 percent of all banded woodcock (figure 12). Michigan and Maine together account for 49 percent of all banded woodcock. Maine is home to Moosehorn National Wildlife Refuge, the only site in the national wildlife refuge system dedicated to woodcock research. Eleven states and provinces banded more than 90 percent of all banded woodcock.

Marking birds with bands has a long history. In Europe, it's called ringing. Lincoln (1921), Cole (1922), and Wood (1945) all give good reviews of this history. The use of metal bands to identify birds goes back at least to the sixteenth century. Falconers during this time banded their birds to identify them if they became lost. In the United States, Audubon was the first person to mark birds by placing a silver cord on the legs of nestling phoebes in 1803. The next year, two of these nestlings returned with their silver cords still intact.

The first systematic or scientific use of banding or ringing started in 1899, when a Danish schoolteacher named Hans Mortensen banded birds with a return address in the hopes that someone would find the bands

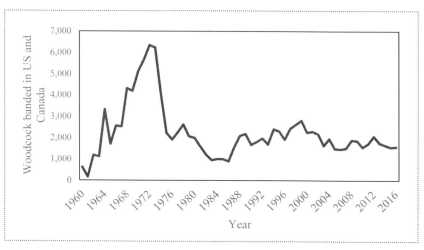

Figure 11. The number of woodcock banded in the United States and Canada from 1960 through 2016. Data retrieved from www.pwrc.usgs.gov/bbl/.

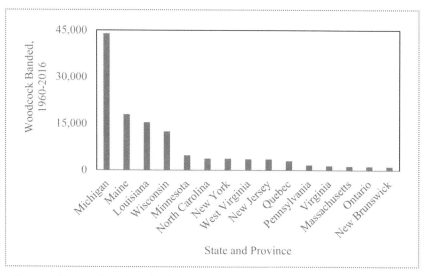

Figure 12. The number of woodcock banded in states and provinces from 1960 through 2016. Data retrieved from www.pwrc.usgs.gov/bbl/.

and report the location back to him. Three years later in the United States, Paul Bartsch banded more than a hundred black-crowned night-herons in Washington, D.C. His bands were inscribed with instructions to return them to the Smithsonian Institution. However, the person who really got banding started in North America was Jack Miner of Ontario, who banded 20,000 Canada geese between 1909 and 1939.

In 1904, Percy Taverner of Michigan initiated the distribution of aluminum bands, creating two hundred bands to give to ornithologists. Taverner had very specific ideas about what types of information scientists needed and could get from bands that were found and returned, including life spans and migration routes.

In 1909, Leon Cole founded the American Bird Banding Association. This group of dedicated birders and professional ornithologists oversaw banding until 1920, when the federal government took over these efforts following the 1918 Migratory Bird Treaty Act among Canada, the United States, and Mexico.

Frederick Lincoln was put in charge of organizing these efforts for the U.S. Biological Survey, which later became the U.S. Fish and Wildlife Service. Lincoln ran the program from 1920 to 1946 and developed systems for organizing, cataloging, and eventually analyzing the results of the banding program.

Today, bird bands come in a range of sizes. Each band contains a unique nine-digit number. People finding a banded bird who contact the Bird Banding Laboratory in Patuxent will receive a notice telling them where the bird was banded and when. Bands come in sizes X to 9C, ranging from smallest to largest. For some perspective on what these sizes mean, size X fits a hummingbird, 0 fits most warblers, 1 fits most sparrows, 2 a red-headed woodpecker, 4 a pileated woodpecker, 5 a wood duck, 6 a gadwall, 7 a mallard, 8 a great horned owl, and 9 a trumpeter swan. Woodcock, along with mourning doves, flickers, and blue jays, receive a size 3. Bird banding is very regulated. Not just anyone can order bands. Banders must be well trained, certified, and permitted with the Bird Banding Laboratory.

Mendall (1938) highlights several other methods for capturing and banding woodcock. The first is night-lighting birds with high-powered flashlights. The birds are stunned by the bright lights and can be scooped up in a net. He cites earlier studies describing how to erect a fence around a nesting woodcock to capture the nestlings before they can wander off. This seems like it would be quite a disturbance for the nesting hen and would alert predators in the area to something that might be worth investigating.

Norris et al. (1940) and Sheldon (1955) describe a bow trap placed on the singing grounds, usually with a decoy hen in the middle. When a male tries to mate with the decoy, he triggers the trap, and a net tied to a wooden bow closes over him.

Today, mist netting for woodcock is a common capture technique. Large, very finely threaded nets with horizontal strings running through them are placed around the singing ground. Either the male or a visiting female hits the net, falls down into the pockets created from the net hanging loosely on the horizontal lines, and becomes entangled. The researcher then gets the job of untangling a long-billed bird with big feet from a black net in the dark.

As testimony to the amazing vision of these birds, I once watched a hen woodcock approach a netted singing ground after sundown. The black net was set against a dark background, and no wind was blowing the net to give its location away. The hen, flying directly at the net, pulled up just a couple feet from it. Without seeming to slow down, she changed directions instantly, flew straight up over the net, and continued onto the singing ground. It was amazing to see both how good her vision was and how quickly she could change directions and maneuver over the net.

Bird banding is one of the most exciting parts of wildlife management, especially for students and beginning scientists. It's one of the few times biologists get to touch and handle the birds they are studying. Biologists can sometimes go for weeks without actually seeing any study animals, or they see them only through binoculars from a distance. Often, biologists just study signs that the animals have been in a particular location. These days, with satellite tracking and automatic data downloads, biologists don't have to leave the office.

That said, handling animals can be very invasive, potentially cause them injury, increase their stress levels, and affect their use of a particular habitat or their long-term behaviors. Every researcher is cognizant of these factors, and all want to minimize them. The purpose of research is to study how the animal naturally behaves. Any way the researcher affects those behaviors will have a negative impact on the research. While handling an animal is often necessary, the goal is always to do it with minimum disruption to the animal, to other animals such as the hen if banding her young, or to the habitat.

Banding is fun, but it has a serious purpose also. Banding data are one of the most powerful tools biologists have, and these data become even more powerful when combined with other types of data. Most easily, banding can tell us about migration patterns; since we know where the bird was banded and where it was recovered, we can estimate the path it took to get there. When a bird band is recovered within the same year as the bird was banded, it's called a direct recovery. Bird bands returned more than a year later are referred to as indirect recoveries.

We can also make estimates of the longevity of individual birds as well as develop life tables for the population. Life tables, actuarial tables in human terms, tell us how many individuals survive to different ages. The oldest American woodcock known was a female banded as a hatch-year bird in Wisconsin on June 30, 1971, and harvested in Wisconsin more than eleven years later on October 2, 1982. However, the majority of birds never see their first birthday.

Krohn and Clark (1977) examined 639 bands recovered from woodcock in Maine between 1962 to 1974. That sounds like a lot, but researchers have to band 9,643 birds to get that many bands back. In other words, there is a 6.6 percent recovery rate. In Michigan, Krementz et al. (2003) reported a recovery rate of only 3.6 percent. Birds that aren't hunted have a much lower return rate. As Aldo Leopold said, "To band a bird is to hold a ticket in a great lottery" (1949).

Krohn and Clark found that the bands in their study were recovered over an extensive area from New Jersey to Texas, with most concentrated from Virginia to Florida. For birds harvested in Maine, a large percentage of the bands were recovered relatively close to where the bird was first banded. They concluded that there are two relatively distinct flyways. Birds from New England and the Maritime Provinces overwinter in the southeastern states, while birds from the Great Lakes area overwinter in Louisiana. Krohn and Clark do point to the need for more banding efforts between the Great Lakes states and New England to determine whether these are distinct flyways or just an east-west gradient.

Krementz et al. (2003) noted that 90 percent of the birds banded in Michigan were recovered in Michigan. Another 3 percent were recovered in Louisiana, and the remaining were scattered among eleven other states. They found that survival rates of adults were almost twice as high as those of juveniles. Dwyer and Nichols (1982) found similar results but also found that female survivorship was higher than that of males. Survival of young males was especially low.

Surprisingly, several studies found female survivorship to be greater than male. In many species, female survivorship is lower because females are vulnerable when sitting on the nest and raising their broods. The researchers speculate that the higher mortality rates may affect males returning to their breeding grounds early and getting caught by late-winter storms. Alternatively, the sky dance is done for the attention of female woodcock. But the dance also attracts people and, maybe, the occasional predator.

> The woodcock may be found by those who seek him and know his haunts, but it is only for a short time during the breeding season, that he comes out into the open and makes himself conspicuous. (Bent 1962)

6 migration

At the time when the Woodcocks are travelling from the south toward all parts of the United States, on their way to their breeding places, these birds, although they migrate singly, follow each other with such rapidity, that they might be said to arrive in flocks, the one coming directly in the wake of the other. This is particularly observable by a person standing on the eastern banks of the Mississippi or the Ohio, in the evening dusk, from the middle of March to that of April, when almost every instant there whizzes past him a Woodcock, with a velocity equalling that of our swiftest birds.—*Audubon 1840*

These birds, however, like the snipe, are very uncertain in their movements, being governed a good deal by the state of the weather and the consequent condition of the soil.—*Lewis 1906*

While robins bask in the South, woodcocks waft north by night until, backlit against the dawn, they flutter mothlike into brushy draws.—*Williams 2004*

Here today and gone tomorrow, these harbingers of fall are as fleeting as frost on a pumpkin and as ephemeral as wood smoke curling from a chimney on an October morning.
—*Osborn 2016*

Every bird-watcher anxiously waits through the frozen winter for spring migration. The feeders in the yard are overflowing with sparrows. Brightly colored warblers balance on the tip of every branch singing a song for all the world to hear. Every bird hunter anxiously waits through the sweltering summer for birds to return from the north. The honking of Canada geese can be heard every evening through open bedroom windows, announcing their presence for miles around.

Shorebirds are some of the world's champion migrators. Some of the mass flights of birds are so large that biologists use weather radar to study them. They create what is basically a storm front of birds. The white-rumped sandpiper starts at the Magdalen Islands in the gulf of the Saint Lawrence

River and flies nonstop 2,900 miles to the coast of Suriname in South America. With no tailwind assistance, at a flight speed of 40 to 45 miles per hour, it takes these birds 65 to 75 hours of nonstop flight to reach South America. With a strong tailwind, that time is cut significantly. Several shorebirds fly nonstop from Alaska to Hawai'i, a trip of almost 2,500 miles.

However, the current record setter is the bar-tailed godwit, which flies from Alaska to New Zealand without stopping, a trip of more than 6,800 miles lasting eight days. The physiological stress of flying that far is similar to that of a human sprinter, if not more so. Imagine running at a sprinter's pace for eight days without stopping, eating, or drinking.

The Evolution of Migration

One of the first questions we must ask is why all birds, not just woodcock, migrate. It's easy to understand why birds fly south in the fall. It's cold up north. For probers like woodcock, the ground is frozen. If that's the case, why not just stay south all year long? Why migrate north in the spring?

This question isn't settled in the scientific literature, and we frequently see new research papers with new data and new theories for the evolution of migration in general and for individual species or groups of species. Migration is still a very active area of research.

Migration presents several dilemmas. Birds must launch themselves into the evening sky, flying south all night or longer, hoping that when they land in the morning they will be able to find a spot that provides both cover and enough food to refuel for the next leg of the journey. And they have to hope that they can find those spots at dozens of places along the flyway. They've always had to dodge predators such as owls along the way. In modern times, they have to dodge obstacles such as buildings and utility lines.

While many birds, especially most other shorebirds, migrate in large groups, woodcock migrate singly, and thus young-of-the-year birds have no one to follow. How do four- or five-month-old birds know that some night in the fall, they should point southward and fly in a direction they have never flown to a place they have never seen? I'm not sure what I was doing by the time I was eleven months old, but I'm pretty sure I hadn't made a transcontinental solo trip and back by that age.

This gets at the question of how migration evolved. If southern climes are good enough for the winter, why aren't they good enough for the summer? We know that some woodcock breed in the Gulf coast states. Why go through an arduous continent-stretching flight full of uncounted risks?

Kerlinger (1995) argues that the foundation of all explanations of migration has to be food. If birds can have greater survival and reproductive success by moving instead of staying, then evolution will push them to move. While no one bird is smart enough to do the math, evolution simply calculates the costs and benefits over generations and centuries.

The shorebirds that migrate to the High Arctic supply some of the best examples of this principle. First, birds can spread out over all of Arctic Canada and Alaska and even into Russia or Greenland. That's a lot of room. Second, in the land of the midnight sun, birds can forage for food almost around the clock during the middle of the summer. Last, the Arctic is known for the amount of insects it produces. Thus, shorebirds migrate because they have more room to spread out, more time to forage, and more to forage on.

Woodcock Migration via Anecdotal Observations

There's a lot of conventional wisdom around migration. It's said that spring migration is more predictable, and birds may use more predictable cues like day length to begin migration. The move northward is a race. Whoever gets there first gets to stake out the best habitat. Whoever has the best habitat will attract the most females. Females who are in the best habitat will have a higher probability of successfully raising broods. If you show up late on the northern grounds, you could end up in second-rate habitat with second-rate breeding success.

With their short, rounded wings, woodcock simply aren't built for long,

sustained flights like other shorebirds. There are some unusual occur-
rences, though. Rosene (1949) records a woodcock that landed on the flight
deck of the aircraft carrier USS *Midway* when the carrier was 108 miles away
from the nearest point of land. What's especially interesting about this
record is that it occurred on July 6, well before the migration period. Pettingill (1936) lists hypothetical records of woodcock on Bermuda in October
1859 and Jamaica in November 1864. Bermuda is almost 1,800 miles from
the U.S. mainland. All three of these birds would have had a very long flight
on wings not designed for that type of travel.

If these were true sightings, we must assume that these birds were
caught by a violent storm and blown to these islands. A quick look at a map
shows how improbable it is to randomly get blown to Bermuda, a tiny speck
in a vast ocean. For every woodcock or other bird that ends up on an island,
how many find themselves eventually settling down, exhausted, starving,
and dehydrated, into the ocean waves never to be seen again?

Although many birds fly at relatively high altitudes during migration,
where the cooler, thinner air makes it easier to thermoregulate and provides less resistance, woodcock migrate at relatively low levels. Pettingill
(1936) and Knight (1946) both estimate that they fly about 30 miles per night
during migration. Sheldon (1967) states that they migrate at rates between
6.5 to 53.6 miles per night. Using radiotelemetry data, Myatt and Krementz
(2007b) conclude that they migrate at a rate of 23 to 55 miles per night.
Moore (2016) found that they travel an average of 43 miles per night in the
fall and 25 miles per night in the spring.

Woodcock may be some of the slowest fliers in the bird world. Mendall
and Aldous (1943) note that some birds fly as slow as 5 miles per hour; they
cite several other studies that describe the flight speed as 13 to 36 miles
per hour. These speeds were calculated by researchers driving a car while
watching the road in front, craning their necks out the window trying to
keep track of a flying bird, and checking the speedometer. It is amazing that
wildlife biologists survived to retirement back in the day. This is definitely
not a method recommended for today's graduate students.

Woodcock are some of the earliest of spring migrants. It's not unusual to see them on their northern breeding grounds by early to mid-March. Those earliest migrating males are tempting fate. The first to arrive get the best habitat. The first to arrive are also the most susceptible to spring snowstorms, which aren't all that rare across the Great Lakes and New England regions. Wishart and Bider (1976) report that woodcock arrived near Montreal on March 29, 1972, when the snow was still twelve inches deep; in the Laurentians, they didn't arrive until April 20, but the snow was still twenty-four inches deep. Thus the constant push and pull of evolution.

On the other hand, fall migration can be a more leisurely affair. There's really nothing to get to the southern wintering grounds for.

Woodcock seem almost as reluctant to leave their northern haunts in the fall as they are eager to reach them in the spring. (Pettingill 1936)

Fall flights appear to be more irregular and even more dependent on weather conditions than those of spring. (Mendall and Aldous 1943)

In most cases, woodcock fly south in a leisurely fashion, stopping two or three days in one latitude. (Sheldon 1967)

Woodcock often approach the idea of migrating as if it's a notion that just occurred to them.

They'll hang around in coverts for days on end, loafing and doodling as if they have no intention of seeing the next country. Then the wind shifts or stars collide or something happens that only a woodcock can see, and an hour later they're all gone. (McIntosh 1997)

Best to stay up north as long as the food and weather hold out. Then you can mosey along south. There's no big rush. During mild falls, woodcock may trickle south all season long. Alternatively, if a large storm moves through, the birds may catch a tailwind, and large numbers may move out over a short period of time. Peak flights at Cape May, New Jersey, occurred as early as October 18 to 27 in 1923 to as late as December 8 to 14 in 1910.

Some years had no peaks (Pettingill 1936).

> If anything sets my gypsy blood astir it is the combination of flaming October leaves and woodcock. The birds themselves are like gypsies on the move ahead of fall storms, eventual destination South but unable to resist prolonged bivouacs along the way. (Evans 1971)

> I defy anyone to choose a single date and declare that woodcock will be plentiful then. Trickle-through or flight, they will be there, or they will not. . . . The whistledoodle comes and goes when he pleases. (Woolner 1974)

> Paul grinned. "Woodcock," he said. "There's a strange little sprite. They're here when you find them."
>   Which is exactly what I wanted to hear. Any guide who guarantees woodcock is pulling your leg. (McIntosh 1997)

As Mark Twain supposedly said, "Never let the truth get in the way of a good story." The preceding descriptions were the conventional wisdom about woodcock migration. But there's nothing conventional about timberdoodles.

There are dozens of observations saying that the timing of fall woodcock migration is somewhat unpredictable and that different sexes or ages of birds migrate at different times. However, the data may not support these ideas. Coon et al. (1976), Sepik and Derleth (1993), and Meunier et al. (2008) found no difference in age or sex when it came to time of departure.

Myatt and Krementz (2007a) studied banding results from 1929 to 2001 and wingbee survey results from 1963 to 2003 from Minnesota, Wisconsin, and Michigan. Looking at information from two different datasets, both stretching over multiple decades, they found that the majority of birds didn't move south until the first half of November. Meunier et al. (2008) showed that there was no strong correlation between migration and weather factors. The strongest variable in their model was day length, which doesn't vary from year to year.

The apparent absence of temperature dependent migration . . . suggests
that migration is fairly constant among years.
(Krementz, Seginak, and Pendleton 1994)

. . . migration tended not to vary among years. (Myatt and Krementz 2007b)

Despite varying local climate conditions, woodcock in our study initiated
migration around the same time each year. (Meunier et al. 2008)

So how do we explain this discrepancy between contemporary data and
so many anecdotal observations over the decades? One idea is that woodcock born north of the Great Lakes states move through first and are then
followed by Great Lakes birds. Second, birds may be moving around significantly in the weeks before they migrate. This is a condition called Zugunruhe, German for "migratory restlessness." This may cause local birds to
appear or disappear seemingly at random. Coon et al. (1976) found that
woodcock with radio transmitters stayed within half a mile of their capture
site until about two weeks before they began migration. Then more than
half the birds made movements of half a mile to five miles.

This might be an extreme case of Zugunruhe or just a young bird who
lost its way, but a one-day-old chick banded at Wheeler National Wildlife Refuge near Decatur, Alabama, on March 3, 1976, was harvested by a
hunter on October 1 of the same year in Midland County, Michigan (Causey
et al. 1979). The two locations are roughly 650 miles apart and generally in
the opposite direction we expect of birds flying in late summer and fall.

The last myth that some of the data may dispel is the rate of migration in
spring and fall. Although the birds should race north for the best breeding
territory and mosey slowly to the south, Moore (2016) found that their fall
migration lasted on average thirty-one days, while their spring migration
was almost twice as long at fifty-three days.

In addition to when the birds migrate is the question of where they
migrate. On the migration path, they can end up in some unusual places.
These are often wherever the birds find themselves when their tired mus-

cles greet the sunrise. Peterson (2006) recounts a day when a woodcock landed on the ledge of a fourth-floor window in downtown New York City.

Woodcock are sometimes in the most unexpected places. (Vale 1936)

When a flight is on, look for John [Woodcock] wherever you find him. He will stop most any place he is tired, and daylight tells him when it is time to rest. (Holland 1944)

Who knows where to look for woodcock? Their haunts are so varied that one may not be surprised to find them almost anywhere, especially on migration. (Bent 1962)

"Woodcock," he said solemnly, "are where you find 'em." (Tappley 2000)

In the end, all we really know is that the woodcock are in those places they are, they are not in the places when they are not, and we don't know exactly why. (Keer 2016)

Woodcock also key in on some really small features of the landscape during migration. There are numerous reports in the literature stating that hunters can expect to find them not only in a particular area, but they can often flush them from the exact same location where they had previously flushed them.

Never overlook a spot where you have killed birds on previous hunts. Year after year, birds will be found not only in the same cover, but in the same spot in the same covers. You can kill a single bird from such a spot and find another there the very next day and day after, when the birds are moving through. (Holland 1944)

It is interesting to observe how often certain particular spots in a woodcock cover continue to attract birds year after year, or for that matter, day after day. By "spots" I mean exactly that. Not areas of half an acre or so, but "spots," frequently not more than a square yard or two in extent. . . . Some men have told me of killing a dozen or more birds in a single season, all flushed from the same identical spot. (Knight 1946)

Migratory or not, a given woodcock that runs into a charge of No. 8 is likely to be replaced immediately by another bird in almost exactly the same spot, often within only a few days, a ghostly process unexplained by topnotch hunters. (Waterman 1972)

This jewel may be no more than a stunted apple tree at the edge of a birch hillside, a tiny swirl of grapevines at the edge of a swamp, a few popples on a knoll, or an edge where the highbush blueberries fight with the juniper for living space. If a bird has been there, rest assured that another will arrive. . . . Somehow a migrating 'doodle knows that this is a feeding ground, and flutters in. Find a hot little corner and visit it regularly; you won't be disappointed. (Woolner 1974)

As woodcock migrate, they may stop along the way to recuperate from the stresses of the trip. Myatt and Krementz (2007b), using radio transmitters on birds, found that stopover duration at particular locations along the migration routes ranged from one to fourteen days. Interestingly, they observed that the birds were occupying mature oak forests. Moore (2016) found the average minimum duration at a stopover was 5.1 days in the fall and 5.9 days in the spring.

These were birds originally captured in Minnesota, Wisconsin, and Michigan. Birds flying down the Mississippi River valley have to fly over what is essentially the Corn Belt, a wide area across Iowa, Illinois, and Indiana, as well as other states that may not provide an abundance of ideal woodcock habitat. The results might be different if the same study was done in the East, with its larger tracts of continuous forest.

For many shorebirds that migrate in large flocks, it's easy to identify focal areas on the landscape critical to these species. Cheyenne Bottoms in Kansas is a large wetland complex that is a crucial stopover for hundreds of thousands or even millions of birds. We know the site and can protect it. However, woodcock migrate independently and land wherever they end up in the morning. Therefore, it's hard to identify and even harder to protect any crucial habitat along their migration paths. However, Myatt and Krementz (2007b) were able to identify several general areas where a large percentage of the birds in their study stopped.

Given all those variables and after traveling the length and breadth of the continent each spring and fall, woodcock often end up exactly where they started. This does make sense. If you are born and raised in an area, it's almost by definition a good place to nest and raise your own broods. If it's an area where you survived the winter last year, it's a good guess that you should return there, as you can probably count on finding enough food and habitat again this year.

Mendall and Aldous (1943) captured a male on a singing ground a year after he was born. His singing ground was only four hundred yards from the nest where he hatched the previous year. Dwyer et al. (1982) found that 20 to 50 percent of the nesting hens they captured in their study at Moosehorn National Wildlife Refuge had been banded in the study area in a previous year. Five of these females were captured and banded as chicks in the area previously. Two of those five hens were caught within a few yards of where they were banded as chicks.

Sheldon (1967) recorded distances between singing grounds of individual males over two successive years. Thirty-two percent returned to the same singing ground or to a very near one. Sixty-three percent were recaptured the following year on a singing ground within half a mile of the ground they were banded on the previous year. Among adult females, eight of eleven were recaptured one to four years later in the same summer field where they were banded.

Sheldon found the same trend at the other end of the birds' flyway. Seventy-one percent of woodcock banded in fields in Louisiana were recaptured in following years in those same fields. The numbers weren't quite as high in their North Carolina study, but Connors and Doerr (1982) recaptured a small percentage of the birds they banded in subsequent years, often in the same field where they were originally banded.

What all this tells us is that woodcock have very good maps in their heads, allowing them to find places they haven't seen for months. The fact that different birds find the same site, often within feet or yards of the last bird, tells us that these birds are seeing the landscape on a far finer scale than we are. Our models and management are done at the scale of acres or larger. These birds may be looking at the same habitat one square foot at a time.

Today, we divide woodcock into two management regions, the central and the eastern. The line dividing them runs down the eastern edge of Ohio, Kentucky, Tennessee, and Alabama. Mendall and Aldous (1943) state that "the various routes used by migrating woodcock still remain somewhat of a mystery." They go on to point out that birds recovered in Louisiana and Mississippi were originally banded anywhere from Minnesota to Maine. Knight (1946) speculates that woodcock use three flight lanes: the Eastern Seaboard, another lane immediately west of the Appalachian Mountains, and the Mississippi River valley. Sheldon (1967) also identified Atlantic, central, and western routes.

Sheldon's Atlantic route runs east of the Appalachian Mountains with birds overwintering across the southeast from Louisiana to Georgia and the Carolinas. The central route starts in Louisiana and goes north into New York and Quebec. The western route also starts in Louisiana and eastern Texas; these birds spend the summer in Minnesota and Wisconsin. Knight (1946) identifies three similar routes but writes that "boundary lines are quite vague if they exist at all." Myatt and Krementz (2007a) and Moore (2016) also show the possibility of three different routes but again point to a lot of overlap and fuzziness between borders of these routes.

Krohn and Clark (1977) say that banding returns from Maine and Louisiana indicate two relatively distinct flyways. However, they hypothesize that as more data are collected, "bandings will show that flyways are not distinct units but that there is an east–west gradient in breeding-wintering ground relationships." Sullins et al. (2016) concluded from their study using isotopes that organization by the Central and Eastern Management Regions is justified.

Woodcock summer across the Great Lakes region to New England and into southeastern Canada. Evolution is usually measured over hundreds of thousands to millions of years. However, just ten thousand or so years ago, much of the woodcock's breeding range was covered under a mile-thick sheet of ice, a glacier. In evolutionary time, the last glacier retreated a few blinks ago. How did this behavior evolve relatively quickly?

Rhymer et al. (2005) studied the genetics of woodcock in the central and eastern regions and found that none of the genetic variability can be attributed to differences between regions or among populations. They go on to say that the genetic data indicate that woodcock have undergone a recent postglacial expansion into their modern breeding range. Interestingly, they found that genes flow primarily from the eastern to the central region with little movement of genes (and birds) from the central region to the eastern one.

All these data show the power of new techniques and technologies in science. For decades, scientists were limited to one method for determining flyways: banding. Today, they have long-distance radiotelemetry and satellite tracking, genetics, and isotopic analysis. Each is powerful alone, but they become more powerful when multiple techniques are combined in analyses.

One area where folklore and facts still haven't come to a final conclusion concerns the role of moonlight in the migration of woodcock and of all birds. It makes sense that moonlight would make it easier to see during night flights. Parman (2015) gives a very good review of the hunting and research literature on moonlight and migration. Meunier et al. (2008) are the most recent researchers to study this issue. They found that moon phase was an important predictor of departure, with woodcock preferring to depart southward in the fall under the gibbous moon's greater than 50 percent illumination.

> I still envision timberdoodles etched against a full moon, even though I know this is ridiculous—but is there any man in our company who wars against a dream. (Parman 2015)

Even if the science still leaves some questions about moonlight and migration, on a crisp evening in spring or fall while we take the dog for a last walk before bedtime, the romantic in any of us can't help but think about migrating birds.

7 thinking about dynamic landscapes

One bad feature about a good birch-aspen cover is that it won't stay good for any length of time. While the young trees are in the whip or sapling stage, the woodcock take to them like ducks to water. As soon, however, as these whips have grown into trees thirty or forty feet high—a matter of relatively few years—the woodcock no longer use the cover. —*Knight 1946*

The abandonment of submarginal farm land in parts of the northern woodcock range in recent decades improved some habitat as cropland converted to grass and brush. This land too is rapidly growing into forest, and therefore the benefit is temporary.—*Edminster 1954*

The woodcock, like the quail, found his range in eastern North America much extended by the destruction of the original heavy forest growth. The gloomy depths of the woods had no more appeal to him than the plains west of the Mississippi. . . . I believe there is more genuine woodcock cover in the eastern states today than ever before.—*Sheldon 1993*

Most woodcock habitat, of course, has succumbed to a different kind of progress. Those pastures and hillsides where head-high saplings once grew have matured into forests, devoid of their thick understory that woodcock need for survival.—*Tappley 2000*

Gary Larson titled one of his Far Side books *Wildlife Preserves*. The cover shows wildlife contained in sealed canning jars, preserved as we preserve fruits and vegetables from the garden and orchard, with tourists driving through fields of these jars. While this highlights Larson's typical playful use of the double meaning of words, for wildlife biologists and society in general it brings up some larger issues and common perceptions and misperceptions.

Almost by definition, when we preserve something, it doesn't change. We preserve food so it doesn't rot. We preserve items in archives and historical museums so they won't change. However, in nature, nothing stays unchanged even for a short time.

The myth of the undisturbed presettlement landscape is about as ingrained as any other in our collective consciousness. First, this ignores the role that Native Americans played in managing, often extensively, their landscape. Second, it ignores the dynamics of nature itself.

In his biography of John Adams, a book which has nothing to do with the natural world or woodcock, John Ferling begins, "At one time trees had stood as far as one could see. Oaks, elm, maple, and a score of other varieties had marched through the valleys and up the slopes of the nearby Blue Hills, their silent trek interrupted only by occasional blue-green ponds and the rivers that had coursed and eddied towards the neighboring coast" (1992). That all sounds very peaceful, bucolic, and static. The next sentence is, "Then the English settlers had come." Cue dramatic music.

In 1963, Starker Leopold, one of Aldo's sons, chaired an advisory board to investigate the state of the national parks. The product of these efforts was a report titled "Wildlife Management in the National Parks," most commonly referred to as the "Leopold Report." The phrase most often cited and remembered from this report is that the parks should be a "vignette of primitive America." A vignette provides a very stable, unchanging view of the world. While not explicitly stated, this phrase implied to many people that nature was unchanging.

Sixty years earlier, President Theodore Roosevelt made the following statement about the Grand Canyon: "You can not improve it. The ages have been at work on it, and man can only mar it." Basically, anything that humans do in the natural world is ultimately destructive. In the following chapters, I hope to demonstrate that people can manage habitats productively and successfully and that tools such as chain saws and axes aren't necessarily destructive.

Succession and Disturbance

Two ecological processes make forests—and any ecosystem—dynamic. Those are ecological succession and ecological disturbance. Succession is the sequence of different plant species at a site following some disturbance.

Imagine an agricultural field, heavily disturbed on an annual basis by farm equipment, which is abandoned. The first year, it will have primarily annual weeds. A few years later, there will be some perennial grasses and flowers. In a few more years, there will be shrubs, followed by fast-growing and sun-loving trees. Finally, the site will be dominated by shade-tolerant trees such as oaks and maples.

The exact species and order of those species in an area will depend on a number of variables. What seeds are in the soil? What nearby plants provide a seed source? Cottonwood seeds can be dispersed by the wind and travel long distances. Ash and maple trees have wind-dispersed seeds, but they don't travel nearly as far as cottonwood seeds. Species like redcedar and pin cherry produce fruits and seeds that birds eat, then fly some distance and defecate. Species like oak and hickory have large seeds that won't travel very far. They will take longer to get to the area.

While succession is somewhat generally predictable, the details of the species, timing, and what the site ultimately looks like if there are no more disturbances aren't necessarily predictable.

Ecological disturbance works in the opposite direction, resetting the clock on succession. Disturbances can take many forms. They have different severities and sizes. Volcanic eruptions as well as glaciation and glacial retreat are probably some of the greatest disturbances, leaving a virtually abiotic environment. Other disturbances such as forest fires or tornadoes can be devastating but still leave some plants or seeds in the soil. In other cases, a disturbance could be a few trees in a forest falling over in a storm. Disturbances can range in size from the path of a hurricane to a gopher mound.

These two processes, succession and disturbance, operating at different intensities, at different locations on the landscape, and at different frequencies work together to create dynamic plant communities and wildlife habitats across a region. Those dynamics, as well as different starting positions in the process and different plant species at the site and near the site, affect what species are present at any particular location at any particular time.

Pre-Columbian America

Many would argue that young forests are a direct result of deforestation and that deforestation is a modern, human-caused issue. In pre-Columbian America, all the trees were tall and all the forests were continuous. The direct and indirect evidence simply doesn't bear this out. The old adage that a squirrel could leap from treetop to treetop from the Atlantic Ocean to the Mississippi River just isn't true.

There are many natural causes of forest disturbances. Hurricanes occasionally batter New England, the southeast, and the Gulf coast, knocking down trees or entire forests and causing mudslides and other natural disasters.

> [In eastern Texas] the greatest number of woodcock were found in a site having dense pine sapling regeneration in a hurricane blow-down area. (Sullins and Conway 2013)

Tornadoes would have been more common in the Great Lakes region and along the woodcock's migration routes. Windstorms can devastate large swaths of forests. Ice storms can break limbs and topple trees over large areas.

These disturbances would not have affected just small areas of forests. Reporting on a study of the original Great Lakes region land surveys, Loucks (1983) found evidence of blowdowns in 25 percent of the 1,600 townships in Wisconsin. He then calculated that the annual rate of blowdowns in Wisconsin was 10,900 acres. Lorimer (2001) shows this similar pattern over a much larger part of the continent.

Beavers were abundant. They cut down trees, create dams, flood meadows, and then abandon those areas after a few years. Beavers alone would create a lot of moist soil and young forest habitat. There were and still are occasional outbreaks of native and invasive insects that kill or weaken trees over large areas.

Then we can start combining these effects. An ice storm, tornado, or hurricane breaks off tree limbs, uproots some trees, and stresses all the trees. This may make them more vulnerable to insect infestations. The insects kill many of the trees that were only damaged in the storm. A fire then sweeps through the area, fueled by the large amount of dried wood.

Native Americans probably had a much greater role in structuring the environment than most people think. Their primary tool for landscape modification was fire (Pyne 1982; Stewart 2002). Many of the earliest explorers describe annual fires in many areas. Different tribes burned different areas at different times of the year or with varying frequency for a variety of reasons. Fire was used as a tool for warfare both offensively and defensively. It made winter travel easier and cleared land for agriculture. It could flush game from an area during hunts or, with circle fires, crowd game into a small area, and game would be attracted to the succulent, nutritious regrowth. Pyne states, "The resulting mosaic of anthropogenic fire regimes is as complex as the historical geography of the cultures themselves."

Wildlife abundance also speaks to there being lots of early successional forests on the continent. We know that bison occurred as far east as eastern New York (Hodgson 1994). Bison are grazers and wouldn't find anything to eat if there were only tall mature forests. Likewise, the heath hen, a subspecies of the greater prairie-chicken, was abundant along parts of the Atlantic coast. Heath hens prefer open scrubby habitats, not closed-canopy mature forests. Cronon (1983) cites several sources that say that the earliest colonists often had to import wood from afar. The dark primeval forests were actually quite sunny and open in many areas. We can imagine lots of young forests at the interface of these open areas and older forests.

Euro-American Settlement

We tend to think of settlers cutting down trees and destroying forests, but that wasn't always the case.

Through the mid- to late 1800s, unregulated lumber extraction devastated the nation's forests, polluted rivers, and caused entire mountainsides to erode away. America expanded west and its population grew, from both births and immigration, at an incredible rate. All those people needed homes, mostly made out of wood. All those homes had to be heated, mostly with wood. Fencing for livestock was often made from wood. Railroads expanded even faster than the population in the post–Civil War decades. Railroads ran on wooden ties, and the early locomotives were powered by wood. Ships for both commerce and the military were made almost entirely of wood. It's almost unimaginable to think how much wood was used and how many acres of forests were cut down.

However, settlers were also adding forests to the landscape. While fire was a common tool for Native Americans, once Euro-Americans arrived with their permanent settlements of wooden houses with wooden roofs with stacks of hay to feed livestock through the winter, fire went from being a common tool to being the enemy. Pyne (1982) and Stewart (2002) cite numerous references to the idea that as soon as settlers became established in an area and excluded fire, the open grasslands or savannas quickly became forested.

In Leopold's *A Sand County Almanac*, the chapter immediately before "Sky Dance" is "Bur Oak." Leopold cites Jonathan Carver and John Muir and provides his own observations about this same phenomenon of forests following settlement. He quotes Muir: "As soon as the oak openings were settled, and the farmers had prevented the running grass-fires, the grubs grew up into trees and formed tall thickets so dense it was difficult to walk through them." Sounds like woodcock habitat. Pyne (1982) writes, "The Great American Forest may be more a product of settlement than a victim of it."

So what are we left with? We have natural events such as tornadoes, hurricanes, blowdowns, and ice storms. Add to that insect outbreaks and beaver activity that would have left acres of standing dead trees more vulnerable to blowdowns. Then we can add fires, whether set by lightning or people.

Native Americans had inhabited the landscape for millennia. Then in a matter of a couple centuries, they were pushed off their lands and decimated by diseases. After that, we add Euro-American settlement patterns to this landscape. Timber barons cut their way through the mature forests, leaving young forests and slash piles behind them. Those slash piles created some destructive fires, again resetting the young forests. As settlers reached the prairie-forest border, they halted fire, producing a thick stand of saplings.

We've all read stories from a century or more ago about the fabulous populations of woodcock, ruffed grouse, and other wildlife. In addition to the timber barons, the settlers also cut down a lot of mature forests and created a lot of young forests.

> Lumbering and repeated forest fires have produced a dense growth of scrub oak and pitch pine, and the area supports a liberal supply of wild life of all sorts. (Knight 1946)

> It is probable that the first full lumber harvest in the East in the 19th century greatly increased the extent of woodcock cover. The fires that followed the lumbering extended the useful time of some of this habitat. (Edminster 1954)

A second factor critical to woodcock habitat, especially in New England, was farm abandonment. Tired of fighting the New England climate, hilly terrain, and rocky soils, people in the late 1800s and early 1900s abandoned their farms to move back to the city or move to the more fertile soils of the Midwest.

> The hill farms of the Atlantic highlands so industriously tilled years ago are fast being abandoned by "the hardy peasantry," and the old cider orchards and pastures are growing up to briars and birches and sumach. Just how this tendency will affect the economy of the nation is doubtful, but certainly it is a splendid thing for the woodcock, the grouse, and the quail. (Sheldon 1993)

> The boom time for the woodcock in this country occurred from the turn of the twentieth century throughout the Great Depression of the 1930s. During

those years family farms were abandoned at an ever-increasing rate as the nation turned more and more to city life. As a result, in the emptying countryside, past patches of cleared pasture slowly returned to woodland: thick stands of sprouting aspen clouded the once bald New England hills, and mucky, worm-rich alder brakes clotted the brooks and burns and bottomlands that drain them: Timberdoodle Heaven. (Jones 2002)

Some reports even tie these two factors together.

Much of the land in the [Moosehorn National Wildlife] refuge was clearcut and burned by wildfire about the turn of the century. At the same time, many of the farms that were economically tied to the forest industry were abandoned as the timber supply declined and mechanization increased. (Dwyer et al. 1988)

The Wildlife Management Institute's "Best Management Practices for Woodcock and Associated Bird Species" in the Central Management Region lists five factors that are key to identifying roosting areas: pastures with light to moderate grazing, recent clear-cuts and log landings, newly established tree plantations, revegetated mining areas, and recently abandoned farmland. Under "diurnal feeding habitats," the first thing the guide says to look for is abandoned farmland, especially overtopped apple orchards.

None of these is an indicator of pristine wilderness. In fact, they are all areas that have been pretty significantly altered by humans in the recent past. Therefore, to maintain habitats for woodcock, humans must continue to take an active role in the management of early successional areas.

The take-home message is that the forested landscape of the eastern United States has a long and very complex history that would have created multiple patterns of vegetation at different spatial scales during different time periods. That squirrel would never have made it from the Atlantic to the mighty Mississippi.

Environmentalism and Saving the Forest

In his poem "A Forest Hymn," William Cullen Bryant begins, "The groves were God's first temples." In *My First Summer in the Sierra*, John Muir expands on this by saying, "No wonder the hills and groves were God's first temples, and the more they are cut down and hewn into cathedrals and churches, the farther off and dimmer seems the Lord." The massive tree trunks resemble the pillars in medieval cathedrals. The large limbs spreading overhead are reminiscent of the arches in those same cathedrals. The sunlight filtering through the leaves above creates a dappled pattern not unlike a stained glass window.

When we walk through a mature forest, it can be a reverential experience (figure 13). When we walk through woodcock habitat, we often use language far from appropriate in a cathedral (figure 14).

The transcendentalists and the romantic poets, Henry David Thoreau and Ralph Waldo Emerson being two names many modern Americans would associate with these movements, wished for a natural world unblemished by humanity. In 1864, George Perkins Marsh's revolutionary book *Man and Nature* gave further credence to the idea that humans are a destructive force in nature when he wrote, "Man is everywhere a disturbing agent. Wherever he plants his foot, the harmonies of nature are turned into discord."

The history of environmentalism has been written about extensively from multiple points of view. Without getting too lost in the details, we can roughly break the early forest conservation movement into two schools of thought. The first was the preservationist view, largely advocated by John Muir. The second was the wise-use perspective, advocated by Gifford Pinchot. This debate was carried on politely—and often not politely—from the latter years of the 1800s into the early decades of the 1900s.

In the 1930s, Franklin Roosevelt's Civilian Conservation Corps planted millions of trees across the country for a variety of conservation projects. Planting trees is how we help repair nature.

Figure 13. An older forest of bigtooth aspen. While nice for us to walk through, there isn't much ground cover for birds to hide in.

In the 1940s, two iconic figures set the course for forest management for decades to come. In 1942, the movie *Bambi* hit the theaters. Animals lived peacefully in the forest. Owls, skunks, and rabbits played together. Humans were destructive forces. Fire was evil. The imagery of that movie is chilling still today. Two years later, one of the nation's most endearing and greatest public relations symbol was born, Smokey the Bear, whom many land managers today call Smokey the Bore. The antifire, antiforest attitude Smokey has instilled in the American psyche makes it very difficult to manage young and dynamic habitats.

If fire was bad before . . . More importantly, Smokey told generations of kids that only they could prevent forest fires. Trees are good. Fires damage trees and are therefore bad.

In the 1980s, there was the spotted owl in the Pacific Northwest and tropical deforestation. The environmental movement convinced the public that anyone with a chain saw or a match was against nature.

Even Aldo Leopold (1949) uses this metaphor when he writes, "When some remote ancestor of ours invented the shovel, he became a giver: he could plant a tree. And when the axe was invented, he became a taker: he could chop it down. Whoever owns land has thus assumed, whether he knows it or not, the divine function of creating and destroying plants." "Giver," "plant," and "creating" have positive connotations. "Taker," "chop," and "destroying" are negative words. A few paragraphs later, Leopold takes a much more circumspect view of the ax. We'll get to that later.

People simply don't like trees being cut down, and they really don't like clear-cuts.

Figure 14. A young aspen forest. While almost impossible for us to walk through, it's a well-protected place for woodcock to nest and raise young.

Woodcock woods are the sort that sprout in the first years after a forest fire or, in the absence of fire, after timbering operations such as clear-cuts. That sapling habitat is ephemeral. And manufacturing it is controversial. "Obviously, trying to convince the public that clear-cutting is a good idea is not in vogue right now," allows Dessecker. (Lipske 1997)

People can become emotionally attached to forests. What starts as a forestry or habitat management question—an ecological issue—can quickly develop into decisions based on opinions and emotions—a psychological or sociological issue.

The logging practice of clear-cutting is particularly vulnerable, and, like dead baby seals, makes a convenient and compelling poster child to point up the horrors of our environmental insensitivity. (Fergus 1992)

And yet acceptance of clearcutting to create early successional habitat is not generally favored by the public, many of whom can be outright hostile to it. Part of this is visceral. Turning a forest is not like turning a field. To those unaccustomed to the sights and sounds of falling trees, even a two-acre clearing can evoke a Loraxian apocalypse. (Freeman 2013)

Today, most good environmentalists would proudly call themselves tree huggers.

Daniel Dessecker has a cutting-edge proposal for nature lovers in the eastern United States. Stop hugging all those trees, he urges, and start chopping some down. (Lipske 1997)

It's no wonder people are confused. First we build an environmental movement largely around deforestation. We tell people that forests should be left alone. We send out armies of citizens to plant trees. Now advocates of young forest management come along and say we need to do more burning and more cutting. Is it any wonder that they meet with resistance from the general public, who have been told one thing for decades, often by the same groups that are now saying something different?

That's the nature of wildlife management. We practice one system of management, thinking we're doing the right thing. As we do that, we conduct research to see what effects our management is having on soils, forests, and different wildlife species. Sometimes our research tells us that what we're doing isn't having the intended effects, and we need to do something else.

To accommodate the full range of species in the North Woods and the annual life cycle of many of these species, we need several age classes of trees on the landscape. What ratio and what pattern of these forest types do we want on the landscape? Our decisions will depend on the species of wildlife we want to encourage or discourage, on local and regional economies, and on social values such as what we want the forests we drive by or walk through to look like. In other words, these are not easy decisions to make, and land managers have to incorporate a wide range of information from very different sources to help them make these decisions.

> The tricky business for land management agencies is satisfying a wide spectrum of forest users, including preservationists who cringe at the whine of a chain saw and bird hunters who want to unleash spaniels and pointers in young aspen. "It almost gets down to which forest is more morally right?" says John Bruggink, a wildlife biologist who tracks woodcock population trends for the U.S. Fish and Wildlife Service. His answer? "Well, we need some of all kinds of forests." (Lipske 1997)

Young forest management for woodcock often seems to be focused on aspen. Therefore, we need to study the term "old-growth" in the context of this species.

8 managing habitats

. . . our woodcock number could ebb and flow with the fortunes of the pulpwood market. Woodcock would thus be highly susceptible to changes in the demand for wood products.—*Hale and Gregg 1976*

We conclude that woodcock do not appear to be sharply selective in choice of nesting cover.—*Coon et al. 1976*

The bird can be as unpredictable in his habits as he is in his flight, and the birds are indeed where you find them.—*Mulak 1987*

Since the majority of eastern forests lies in private hands, private landowners play a critical role in replenishing woodcock numbers.—*Dorsey 1990*

Most plants start out as seeds. This is called sexual reproduction because a seed is created when sperm in a pollen grain fertilizes an egg. The seed germinates, becomes a plant, and eventually dies. An annual plant will live for only one growing season, while a perennial plant may live for decades or in some cases centuries. It's easy for us to identify a maple tree. One seed landed on the ground, germinated, and formed one stem or trunk with many branches and twigs. We can readily identify that as a distinct and individual tree.

Some plants also use asexual reproduction, which arguably isn't reproduction. These species have new stems or shoots that come up from the horizontal roots of the parent tree. There are two stems or trunks coming out of the ground that look like two trees. However, they are connected underground by a common root system. Therefore, they are only a single tree, and more importantly they are genetically identical. In botanical terms, each stem is a ramet, while all the stems of that genetically identifiable individual form the genet (Barnes 1966). Many people refer to ramets as suckers.

It's not unusual to see a clump of aspen growing in a field (figure 15). There may be hundreds of stems in each clump. Often these clumps will have a dome shape. The ramet that originated from a seed and is the oldest is the tallest stem in the middle. As the clone continues to produce roots, new ramets form on the edge of the clump. The new ramets are younger and therefore shorter. However, they are still all one organism, a genet. It's fun to point these out to friends. Ask them how many aspen are in that clump. They'll think about it for a while and come up with a number in the hundreds or thousands. No, the answer is one.

Even larger examples exist. Often in the fall, you can see entire hillsides covered in aspen. One clump will all be the same shade of gold; another area will be slightly ahead or behind in color. The hillside could be covered in aspen, but there are only three or four genetically individual plants on the

Figure 15. This aspen clone represents one genetically individual tree, not many trees.

hill, each of those plants with hundreds or thousands of stems.

Aspen trees can take this concept to ridiculous extremes. Arguably, the largest and maybe oldest living organism is a quaking aspen tree in Utah called the Pando clone. This clone is made up of 47,000 stems. It covers 106 acres and weighs an estimated 13 million pounds (Grant 1993).

One problem aspen clones face is that they are shade intolerant. There is plenty of space between individual stems in an aspen clone for an oak or maple or other type of tree to germinate and become established. Often, these trees will grow up through the clone, overtop the clone, and kill parts of the clone by shading them.

The best way to prevent this from happening is to shear or clear-cut the entire clone, aspen stems as well as all the other trees mixed in with it. Most often, this is done in winter. Practically, this is when the ground is frozen, so foresters can use machinery without getting their equipment stuck in the mud. Botanically, plants use photosynthesis to capture energy and then store that in their root systems over the summer. Winter is when they have the maximum amount of energy stored up. They use this stored energy to leaf out and begin growing again the next spring.

The spring after the aspen clone and invading trees have been cut, the oaks, maples, and other trees are dead stumps. The sun-loving aspen clone now has unlimited amounts of sunlight beating down on the soil and a root system with large amounts of energy stored up. The clone can produce tens of thousands of shoots in the first summer. With good growing conditions, aspen clones can reach six to eight feet by the end of the first summer after the site was cut (figures 16 and 17). The old stems, the ramets, are gone, but the genet, the individual tree, lives on and thrives. This sounds counterintuitive, but one of the best ways to help an aspen survive is to cut it down.

Let's circle back to the concept of young or early successional forests. While the individual stems may look young, in fact be only a couple years old, the genetic individual could be hundreds to thousands to hundreds of thousands of years old. When it comes to aspen, "young" forest may be a misnomer.

Figure 16. My constant companion on the Fourth of July the summer after an aspen shearing. Many of the aspen stems are already three to four feet tall.

Aspen forests also make us rethink the ideas of ecological succession. What we're really doing is starting with an area dominated by older stems of aspen and, through disturbance, converting the plant community to one dominated by young stems of aspen. There will be differences. Some of the taller trees will disappear and some of the lower-growing grasses, wildflowers, shrubs, and species such as raspberries will take full advantage of the sunlight. But aspen will still be the dominant species in the area.

Aspen also turns the idea of old-growth forests upside down. We look at a huge oak tree that may be 150 or 200 years old and often call that old growth. A recently clear-cut aspen clone is considered to be a young forest full of spindly stems. However, if that young aspen is thousands of years old and the oak is only 150 years old, either something is mislabeled or

Figure 17. My same companion in the same area at leaf-out the following spring. In the first full growing season, the aspen grew to six to eight feet tall.

someone is doing the math wrong. Very little is what it seems to be in an aspen-dominated forest. That old oak may have shaded one of the last bison herds on a hot day, but that aspen may have sheltered a woolly mammoth at the end of the last Ice Age.

What do today's aspen forests look like? They look mature, and they are getting more mature. Figure 18 shows that across the Great Lakes states, the acres of aspen-birch forests are declining. Additionally, in the northeast, as figure 19 shows, there are few young aspen-birch forests on the landscape compared to the relatively more mature forest classes.

There is no right or wrong shape to these graphs. It's not good or bad to have this number of acres in each age category. However, we do see that there is little young forest on the landscape. There is no ideal mix of forest

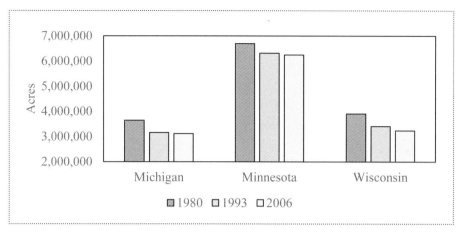

Figure 18. Decline of aspen-birch forests in the Great Lakes states between 1980 and 2006. Adapted from Domke et al. 2008.

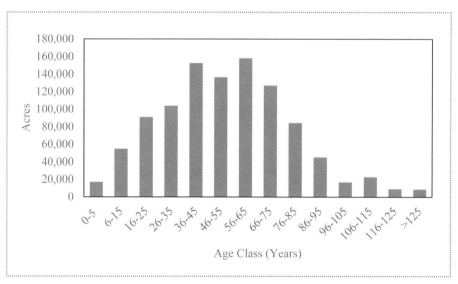

Fig 19. Age structure of aspen-birch forests in northeastern states showing few young forests on the regional landscape. Adapted from Pan et al. 2011.

types or ages on the landscape. We need all types. The ratio of those types is an ecological, economic, and social decision. In other words, it's not an easy decision.

Working Lands

When woodcock hunters and grouse hunters want to find their birds, still today they will often seek out old orchards. Birds seem to be drawn to these areas.

> . . . each of these old farms has its apple orchard. These hardy trees, although they now bear stunted fruit, still flourish, resisting the encroachment of the second growth that claims the land surrounding them. Under their spreading branches there is almost invariably a lush growth of Canada bluegrass. And they all seem to shelter woodcock. (Knight 1946)

> I have often flushed woodcock under apple trees in abandoned orchards where the soil is enriched by several generations of rotting apples. (Sheldon 1967)

> Forgotten apple trees should always be investigated. (Woolner 1974)

Sepik and Dwyer (1982) report that commercial blueberry fields are popular roosting areas for summer woodcock.

Pastures are another area where woodcock can often be found, especially in some of the wetter areas that contain some brush. Ungrazed grasslands are less desirable. The tall grass is simply too thick to move through and blocks the birds' ability to see potential predators. Grazed grass is shorter, and wildlife can use the paths much as the cattle do. Woodcock aren't the only wildlife species that responds this way. Fresh cow pies are a magnet for worms, which in turn attract woodcock.

> Pasture land is used for feeding to a large extent but pastures must be just so in order to attract woodcock. The birds show a remarkable preference for grass that is about four or five inches high. Long grass does not present

enough freedom of movement and cuts down range of vision. Short grass fails to provide enough cover. (Knight 1946)

Where cattle have browsed a thickly grown back pasture, there'll be well-trodden paths through alder and birch and high-bush blueberries. Indiscriminate manuring probably encourages the propagation of worms, and partial grazing creates little openings that are magnets to traveling doodles. (Woolner 1974)

Woodcock need uninhibited walking room and a reasonable chance of seeing what's going on about them. Light or moderate grazing will often open up such covers and attract woodcock. (Kletzly 1976)

Pastures, orchards, blueberry fields are all places where we can find woodcock. None of these speaks to pristine, untouched nature. And then there are forestry practices.

Woodcock will always be primarily forest birds, and the fate of the woodcock will be tied to the vagaries of the timber industry. We will never be able to do enough forest management purely for wildlife habitat manipulation to sustain or increase populations of woodcock and other young forest wildlife species.

Lumbering makes some improvements in woodcock habitat by opening the canopy of the forest and providing herbaceous and brushy cover. (Edminster 1954)

. . . the increase in the commercial harvest of aspen should help to counteract the trend of deteriorating woodcock habitat in aspen-birch forests. (Owen et al. 1977)

Now with the cooperation of the lumber companies and the foresters in this area we're beginning to understand how to manage the forest to the best advantage of the birds. (Fergus 1992)

In a large state or national forest, the land manager may have multiple habitats, multiple land uses, and multiple economic reasons for conducting a management practice such as a clear-cut or a prescribed fire. That manager has control over most or all of the land and can develop short- and long-term management plans for the entire landscape.

Now imagine the same large area broken into scores to hundreds of ten-, twenty-, or forty-acre parcels. Each parcel has a different landowner, and many parcels have homes and other buildings on them. Getting all those people to agree on and coordinate their land management is simply impossible. This is especially true when it comes to prescribed fire.

Managing that fire would also become far more complex. Now, instead of just developing a burn plan for the forest, burn plans have to take into account property lines, homes and other structures, and people who grew up watching *Bambi*. This is the situation that many wildland firefighters face today. They need to deal with homes, outbuildings, propane tanks, stored chemicals, and so on, what they call the wildland-urban interface, WUI.

> People fed up with life in the tumultuous cities of the Northeast and their
> stultifying suburbs were moving into the countryside of rural, interior
> New England, where land was still cheap and building costs not yet exces-
> sive. Inevitably the sites these people chose for their new homes were
> the very places where ruffed grouse and woodcock had found their best
> living: wooded hilltops with brooks running nearby. The former cityfolk
> brought with them fallacious attitudes regarding woodland and wildlife.
> Clear-cutting, which could have brought back the successional stages of
> woodland recovery so necessary to continued woodcock populations, was
> and remains anathema to most flatlanders. (Jones 2002)

Roger Tory Peterson describes one woodcock sky dance that illustrates multiple aspects of woodcock biology and management.

> For several years one woodcock would continue to perform nightly in the
> clearing behind my studio, but as the brush and second growth took over, it
> too disappeared. (Peterson 2006)

First, we see that woodcock can find habitats literally in people's backyards. Second, because people generally don't like logging operations, it can be difficult to convince them to log their lands. We need to admit that forestry can have its ugly periods. Like making an omelet, you have to break a few eggs. This can cause some consternation for the individual landowner. It may cause greater heartburn for the neighbors, especially if they are of a preservationist mind-set or moved to the country to get away from the racket of the city. But in the end, the logging will be worth it.

> The land can look badly scarred after a timber harvest, even one that's professionally planned and carefully conducted. Standing trees can get snapped off or dinged by equipment. . . . Ruts and muddy logging runs leave some parts of the property nearly unrecognizable. This battered appearance can deter nervous landowners from even thinking about logging their properties . . . but it doesn't take long for the woods to recover, for the mud and gouges to disappear. (Primack 2016)

Finding Woodcock

Where can you most easily find woodcock? That's a hard question to answer. Their wintering habitats in the southern states are different from their breeding habitats in the northern states. Those may be different from the habitats they use during migration. Singing ground habitats are different from nesting habitats. Nesting habitats are different from brood-rearing habitats. Daytime habitats are different from nighttime habitats. Like many species of wildlife, woodcock need a little bit of everything over the course of the year. The goal, then, becomes to manage the landscape in different parts of the country to meet all their needs throughout the day, season, and year.

One of the landmark books of American ornithology is *Wild America* by Roger Tory Peterson and James Fisher. In 1953, the two naturalists took a cross-country trip to explore the wilds of America and the birds that live there. Where did they describe seeing woodcock? They found them in downtown Washington, D.C.

Indeed, given all the research on their habitats, woodcock still end up in the oddest places. Giving exact habitat descriptions can cause anyone problems. For singing grounds, we have the following.

That's basically what you look for: a fairly open field near low, wet spots. Even textbooks and wildlife biologists aren't any more helpful than that. (Carney 1993)

The singing ground itself is apt to be most any type of land that is free from trees and brush. Plowed fields, orchards, lawns, fern beds, meadows, hillsides, pastures, almost any sort of open space will serve. (Knight 1946)

Researchers make similar comments about nesting and brood-rearing habitats.

The cover that immediately surrounds the nest is so varied in both type and extent as to discourage correlations. (Mendall and Aldous 1943)

Woodcock broods were found in virtually all types of cover in the study area. (Dwyer et al. 1982)

The same applies the rest of the year.

More frequently than most people imagine, his explorations extend to the hearts of large towns and cities, where trim gardens and broad lawns form attractive hunting grounds. (Sandys and Van Dyke 1924)

It really is amazing to know the wide variety of cover and environment in which woodcock can be found. (Betten 1940)

Woodcock often appear in unexpected places, such as city parks, yards, gardens, orchards, or even lawns. (Bent 1962)

Many cover types are used by the birds during the brood period and throughout the spring and summer. (Wenstrom 1974)

Statistically, we have the best chance of finding woodcock in their ideal habitats. But like many species of wildlife, they can show up almost anywhere. Unfortunately, because many species of wildlife don't read the literature on their own habitats, at times they are liable to do odd things they aren't supposed to do in odd places where they aren't supposed to do them.

More importantly, these quotes are in the management chapter instead of the habitats chapter just to illustrate how many variables managers must take into consideration when doing any habitat project.

Managing Habitats for Woodcock

The first question any person should ask when thinking about managing a parcel of land for woodcock is, What is appropriate for those acres? What is the soil type? What is the hydrology? What is the current vegetation or habitat, and what is the desired vegetation or habitat? Can reasonable management move the area from its current state to a desired state for woodcock? If the answer is no, then the area should be managed for a different group of wildlife species.

> Conservation planning that requires active habitat management is complicated because not all species can be managed for simultaneously. (Masse et al. 2015)

The surrounding landscape also needs to be considered. If there is an abundance of one habitat type in the general area, then it might be most useful for the wildlife to have another habitat type on the particular parcel being managed. Woodcock and all wildlife are probably keying in on larger landscapes than we may have traditionally imagined. Often we focus our activities on a specific parcel: 40, 80, 160 acres, maybe a square mile of 640 acres. However, the birds are probably thinking bigger.

Bennett et al. (1982) conducted habitat management projects at the scale of nine-square-mile areas. When they managed only 25 percent of this landscape—that's still 1,440 acres—they saw little increase in the number of singing males. However, when they clear-cut 50 and 75 percent—2,880 and 4,320 acres respectively—they did see increases in numbers. They saw the greatest increases with the 75 percent clear-cuts.

The best place to manage for woodcock is on previously disturbed lands.

We recommend that habitat management for woodcock be concentrated on previously farmed land. (Owen and Galbraith 1989)

This quote relates to the distribution of earthworms on the landscape. Because earthworms are introduced in the birds' northern breeding grounds, they are often more common in previously farmed or otherwise disturbed areas than in older forests. This gives more credence to the idea of leaving old-growth forests alone. Manage for young forests where medium-aged forests are already on the landscape. Those are lands we know were disturbed within the last few decades.

"Diversity" is a commonly used word in conservation. That can mean a range of things: plant diversity, habitat diversity, structural diversity, and so on. Diversity can also be studied or measured along a range of scales. Some people study diversity within a square-meter plot. Others study it from satellite images that cover large regions.

Earlier in this book, we've seen that woodcock habitats aren't just aspen forests. Almost all descriptions of their habitats may have aspen as a dominant species, but many other plant species are mixed in.

The area where two ecosystems or habitat types come together is called an ecotone. Researchers and managers have known for decades that these areas contain a high diversity of wildlife. There will be species from one habitat type, species from the second habitat type, and species that specialize in the edge area.

In his landmark book, *Game Management*, Aldo Leopold discusses the principle of interspersion. This builds off the concept of edges to break up habitat types into many small patches and arrange them on the landscape. Leopold uses the analogy of a house and a small community. Each house has a kitchen, bathroom, bedroom, and so on. All the needs of the family can be met within the house. Imagine a small community where one house is all kitchen, a second all bedrooms, and so on. The analogy is to large blocks of habitat. People would have to run from house to house multiple times to meet their needs throughout the day. This wouldn't be convenient, to say the least.

Leopold relates interspersion to edge effects using an imaginary quail covey with a circular home range. Imagine a square mile of land broken into four equal square parts: woodland, grassland, brush, and cultivated field. If we assume that quail need all these cover types over the year, there can be only one covey of quail in the center of that square mile. That covey would use only a small percentage of that area. However, if those four large blocks were broken into numerous smaller blocks and scattered around, then there would be numerous places within that square mile where a covey of quail could survive. Just by rearranging the habitats on the landscape, Leopold hypothesized that the same acreage could hold six coveys instead of one.

Throughout much of the twentieth century, most of our habitat management consisted of creating edges. However, in the 1980s, edges fell out of favor with some managers and researchers. They found that nest predation or nest parasitism for a number of species was higher near forest edges. There is a large and extensive body of research on this topic.

> . . . increases in rates of nest depredation and parasitism have been well documented near forest edges, and these effects may extend >300 m into the forest. However, edge effects associated with silvicultural practices within forested landscapes are, as a whole, poorly demonstrated. (Vitz and Rodewald 2006)

Managers began to focus on large tracts of habitat that minimized edges. In more recent years, in forested habitats, they have started to look at the data again more closely. There's a difference between the edge on an isolated woodlot surrounded by agricultural fields or housing developments and the edge created by a timber harvest in a largely forested landscape. To simplify, the first edge scenario probably does reduce nesting success. But the second scenario may increase landscape diversity. As scientists continue to collect more data on more species, these theories will continue to change over time.

Every management action or inaction will be beneficial to some species and detrimental to other species, at least temporarily. Woodcock are attracted to young aspen forests. Pileated woodpeckers, wood ducks, and other cavity-nesting species prefer older and larger aspen groves. Creating woodcock habitats may mean removing some woodpecker habitats at a local level. Obviously, this idea can be expanded to entire communities of wildlife. However, we don't have to see this as a zero-sum game. Creating woodcock habitats does not necessarily mean we are destroying habitats for other species.

There is also the concept of landscape diversity. How many different types of habitats are on the landscape? In what ratio or percentage? What are they? How are they arranged? Are there a few large blocks or many small blocks? The arrangements of habitats on the landscape can be almost as important as the cover types themselves.

> The domain of the male woodcock has two distinct parts—a wooded territory and an open-country territory. The wooded territory consists usually of small growths of poplar, alder, birch, or willow and frequently borders upon heavy timber. There is usually a moist area in the vicinity. (Pettingill 1936)

> As forest, shrub, and grass aggregated in the landscape, woodcock abundance declined; as these land covers intermixed, abundance increased. (Thogmartin et al. 2007)

> . . . our results indicate that increasing the amount of early successional forest and decreasing the amount of mature forest while increasing the interspersion of land-cover types are likely to increase counts of singing male woodcock. (Nelson and Andersen 2013)

Yahner (2003) states that when all factors are taken into consideration, "habitat for both early successional and mature forest species can be achieved simultaneously within the same managed forest landscape." Vitz and Rodewald (2006) reaffirm this with their statement that "ultimately, forested landscapes containing a mosaic of successional stages may hold the most conservation promise for mature-forest birds."

Anders et al. (1998) concluded, "Management decisions concerning the post-dispersal habitats of migrants should take into account the trade-offs between creating early successional habitats and reducing forest breeding habitat." This might be interpreted as smaller is better when it comes to clear-cutting. However, according to Chandler et al. (2012), "when it is practical and does not conflict with other forest values, larger patches of early-successional habitat should be favored when outlining silvicultural and other wildlife management plans." They conclude that all their research demonstrating the value of early successional habitat to both young forest and mature forest birds provides "the basis for developing a new paradigm" to account for the needs of all these species at all their life stages across the landscape.

Put another way, there's nothing big or grand about woodcock habitats. There should be a tangle of different species and lots of different plant communities within a relatively small area on previously disturbed acres.

Managing Your Acres for Woodcock

The primary goal of this book is not to give anyone specific advice about managing land. Managing your acres is complicated and site specific. The best strategy is to start by contacting a private lands biologist or someone in a similar position in the U.S. Fish and Wildlife Service or a state wildlife agency. Some conservation organizations as well as university extension services and county conservation offices also have consultants available to help landowners plan habitat management projects.

Landowners can download a number of resources. The Wildlife Management Institute has produced "Best Management Practices for Woodcock and Associated Bird Species" for both the Upper Great Lakes and Central

Appalachian Mountains regions. For those in the southern part of the country, there is "Woodcock in the Southeast: Natural History and Management for Landowners" by D. G. Krementz and J. J. Jackson. Visit websites for the U.S. Fish and Wildlife Service and state conservation and wildlife agencies as well as NGOs such as the Ruffed Grouse Society.

One very helpful tool is an aerial photo of the property you plan to manage. This can give you the big-picture perspective of the entire parcel and surrounding lands. One of the first things to do is carry the aerial photo while walking over the parcel. This will allow your eye to match the patterns on the photo with the actual vegetation on the ground.

In many parts of the country, aerial photos go back to the 1930s. Ideally, you will be able to find photos from each decade through your Natural Resources Conservation Service or Farm Service Agency office. This will allow you to see past land uses and note how quickly vegetation changed in different parts of the property. Be aware that those photos are sometimes taken at different seasons, leaf on or leaf off, and this can affect the patterns the eye picks up.

Next, it can be helpful to draw lines around each patch of habitat and then measure how many different habitats are on the parcel and the rough percentage of each habitat. This can be done by eyeballing the photo as well as by more precise methods using a geographic information system. From there, you're ready to start planning.

One strategy that works in a number of locations is to cut strips fifty to a hundred feet wide through the habitats (Sepik et al. 1983). A number of publications report success with this method. The strips should be located so that the adjacent strip can be cut every five years with a twenty-year return interval to each strip. When possible, orient these strips perpendicular to streams or creeks. In this manner, there will always be some ideal soil conditions within the strip. During dry years, woodcock can move down toward the creek or stream to find soil moist enough to probe. In wet years, the areas near the stream may be flooded and the birds can probe farther away from the creek within the same strip. By having strips that are zero, five, ten, fifteen, and twenty years old all in close proximity, you can provide all the birds' habitat needs within a small area.

You can also decide to clear-cut or selectively cut each management area. In my own habitat projects, I'll cut down almost every aspen I can. If one looks suitable for a wood duck nesting cavity, it will survive. However, I'm reluctant to cut down an oak or a hickory. Both are beautiful in the fall and provide mast for deer and ruffed grouse. The remaining trees left in the middle or along the edge of a cut will also provide perches for songbirds, such as golden-winged warblers.

There is no one size fits all when it comes to developing a management plan for a parcel of land. Walk the land repeatedly in every season of the year. Talk to someone knowledgeable and experienced. Based on your observations and conversations, develop a habitat management plan as well as a time scale. You can do this yourself in consultation with a biologist, or contract with him or her to write the plan. Ideally, some part of the land should be managed annually or at least every few years to maintain that mosaic of different-aged forests. A young forest management plan will probably stretch over several decades.

> Thus, an important question is not necessarily can habitat be managed for early successional bird species, but how soon after management must habitat be manipulated further to ensure adequate early successional habitat is available due to loss via plant succession. (Yahner 2008)

> . . . because shrubland birds generally disappear from clear-cuts within 20 years of logging, continually creating new shrubland habitat is vital to the maintenance of bird populations. (Schlossberg and King 2009)

It's important to remember that woodcock management is often forest management. In many cases, you should be able to sell some of the lumber. You may not be able to pay for the habitat project completely, but you may be able to significantly offset its costs. At the very least, you and a few close friends may be able to get several years of firewood.

After the management starts and the first cuts or prescribed fires are completed, keep walking the land and watching the response of the vegetation and wildlife. If something isn't working or could work better, revisit the plan and alter it. It's a management plan, not the Ten Commandments.

This is the basis for adaptive management or strategic habitat conservation. Conduct some management, monitor the response, determine whether the action had the intended results, and if not, modify the management next time.

> The private landowner and resource manager will have to take the lead in producing more woodcock and other wildlife on each acre of remaining land. There will be no cheering crowds, no pats on the back, and only modest economic return for this work. But, there will be a warm feeling of accomplishment as you watch a woodcock courtship display in a clearing you cut one cool February day. (Sepik et al. 1983)

epilogue

Ask any group of conservationists for their favorite or most important quote from the conservation bible, *A Sand County Almanac*, and you could probably start a pretty good argument. But one of those surely high on the list is:

> I have read many definitions of what is a conservationist, and written not a few myself, but I suspect that the best is written, not with a pen, but with an axe. It is a matter of what one thinks about while chopping, or deciding what to chop. A conservationist is one who is humbly aware that with each stroke he is writing his signature on the face of his land. (Leopold 1949)

Young forest management requires a person to be an artist in addition to being a scientist and a manager. We can make all the plans we want over the summer and draw lines on maps. But once outside in the winter woods, we have many decisions to make about exactly where to start and stop cutting, what to cut and what to leave. There are some good general guiding principles for what we should do. But in the field, it still comes down to individual decisions about exactly what to do, what not to do, and where to do it.

Aldo Leopold himself wrote that "game management is the *art* of making land produce sustained annual crops of wild game" (1933, my emphasis). This is where experience and familiarity with the landscape come into play. The science can be taught in the classroom in a few college semesters. The art can be learned only from walking the land and observing it over the years. It takes a lot of time, experience, and miles walked to even begin to understand how to read the land.

If you need to ask advice of someone, look down. If their cuffs are frayed and their boot leather well worn, they've probably put in the necessary miles.

Discussing ruffed grouse, Aldo Leopold said that in terms of conventional physics, grouse represent only a millionth of the mass or energy of an acre of forest. But if the grouse are missing, "some kind of motive power has been lost."

Woodcock are smaller, representing even less of the mass or energy of the forest. But they possess some of that same motive power. And there's another element overlaid on that. Since grouse are resident birds, we know they are out there. The dog may not find them today, but they are out there. With woodcock, it's always a guessing game whether the flight has arrived yet, is here, or has gone.

Woodcock have been referred to as ghostly because we're always so unsure of where they are and so surprised when we find them. Even when they are here, it's hard to know it. It's not unusual to see grouse walking on the trail in front of us, posed on a tree branch above us, or even flushing from the forest edge when a truck drives by. A person could be standing in a forest full of woodcock and never be aware of any of them. They sit tight, blend in, and disappear.

It's interesting how often the word "mystery" is used when people talk about woodcock.

No matter how auspicious the day and the cover, the woodcock is always a veritable apparition. A grouse rockets from the scrub of sumach and thorn-apple, and I am startled, but not surprised, for I half knew he would be there. But a woodcock always surprises me. . . . His goings and comings are glamorous with mystery. (Sheldon 1993)

If you hunt woodcock in the secret places where they live, you will learn many things, some of which may change your thinking. But in the final analysis, do we want to know everything there is to know about this unusual little bird? I, for one, do not. . . . Some mysteries are best left unanswered. (Huggler 1996)

You may know from their sign and their song that woodcock are there; you just can't see them as they sit on the forest floor because they are such secretive birds, so perfectly camouflaged. They add an air of mystery to special places on the land. (Krementz and Jackson 1999)

I am always surprised to realize how few people have ever seen, or even heard of, woodcock. I guess I shouldn't be. They are mysterious, wraithlike, diffident creatures. They travel on the dark of the moon, and during the day they hide in places where sensible humans rarely set foot. (Tappley 2000)

Woodcock themselves embody autumn's beauty; dressed in maple orange and aspen black, they are of the forest. Unlike its sedentary neighbor the grouse, the restless woodcock's need to migrate lends them a special sense of mystery. (Herwig 2016)

Motive power plus mystery equals woodcock.

bibliography

Abbey, E. 1977. *The Journey Home: Some Words in Defense of the American West*. New York: E. P. Dutton.

Abbott, G. A. 1914. "An Intimate Acquaintance with Woodcocks." *Wilson Bulletin* 26 (1): 1-6.

Aitken, W. W. 1938. "Records of the Woodcock in Iowa." *Wilson Bulletin* 50 (1): 64.

Albert, J. 2017. "A New Chapter for Elk." *Minnesota Conservation Volunteer*, January-February: 8-19.

Ammann, A. 1981. "A Guide to Capturing and Banding American Woodcock Using Pointing Dogs." Coraopolis, Penn.: Ruffed Grouse Society.

Ammann, A. 1982. "Age Determination of American Woodcock Chicks." In T. J. Dwyer and G. L. Storm, eds., *Woodcock Ecology and Management: Papers from the Seventh Woodcock Symposium Held at the Pennsylvania State University, University Park, 28-30 October 1980*, pp. 22-25. Washington, D.C.: U.S. Fish and Wildlife Service.

Anders, A. D., J. Faaborg, and F. R. Thompson III. 1998. "Postfledging Dispersal, Habitat Use, and Home-Range of Juvenile Wood Thrushes." *Auk* 115 (2): 349-358.

Artmann, J. W., and L. D. Schroeder. 1976. "A Technique for Sexing Woodcock by Wing Measurement." *Journal of Wildlife Management* 40 (3): 572-574.

Askins, C. 1931. *Game Bird Shooting*. New York: Macmillan.

Askins, R. A. 2001. "Sustaining Biological Diversity in Early Successional Communities: The Challenge of Managing Unpopular Habitats." *Wildlife Society Bulletin* 29 (2): 407-412.

Audubon, J. J. 1840. *The Birds of America*. New York: John James Audubon.

Baicich, P. J., and C. J. O. Harrison. 1997. *A Guide to Nests, Eggs, and Nestlings of North American Birds*. San Diego: Academic Press.

Bakeless, J. 1961. *America as Seen by Its First Explorers: The Eyes of Discovery*. Mineola, N.Y.: Dover.

Baker, R. H., and C. C. Newman. 1942. "Speed of a Woodcock." *Auk* 59 (3): 442.

Barnes, B. V. 1966. "The Clonal Growth Habit of American Aspens." *Ecology* 47 (3): 439-447.

Bennett, C. L., D. L. Rabe, and H. H. Prince. 1982. "Response of Several Game Species, with Emphasis on Woodcock, to Extensive Habitat Manipulations." In T. J. Dwyer and G. L. Storm, eds., *Woodcock Ecology and Management: Papers from the Seventh Woodcock Symposium Held at the Pennsylvania State University, University Park, 28–30 October 1980*, pp. 97-105. Washington, D.C.: U.S. Fish and Wildlife Service.

Bent, A. C. 1962. *Life Histories of North American Shorebirds*. Part 1. New York: Dover.

Berdeen, J. B., and D. G. Krementz. 1998. "The Use of Fields at Night by Wintering American Woodcock." *Journal of Wildlife Management* 62 (3): 939-947.

Berry, C. B., W. C. Conway, R. M. Whiting, Jr., and J. P. Duguay. 2010. "Diurnal Microhabitat Use by American Woodcock Wintering in East Texas." In C. A. Stewart and V. E. Frawley, eds., *The Proceedings of the Tenth American Woodcock Symposium, 2006*, pp. 63-75. Lansing: Michigan Department of Natural Resources and Environment.

Betten, H. L. 1940. *Upland Game Shooting*. New York: Alfred A. Knopf.

Birkhead, T. 2016. *The Most Perfect Thing: Inside (and Outside) a Bird's Egg*. New York: Bloomsbury.

Blackman, E. B., C. S. DePerno, R. W. Heiniger, M. J. Krachey, C. E. Moorman, and M. N. Peterson. 2012. "Effects of Crop Field Characteristics on Nocturnal Winter Use by American Woodcock." *Journal of Wildlife Management* 76 (3): 528-533.

Blackman, E. B., C. S. DePerno, C. E. Moorman, and M. N. Peterson. 2013. "Use of Crop Fields and Forest by Wintering American Woodcock." *Southeastern Naturalist* 12 (1): 85-92.

Bourgeois, A. 1977. "Quantitative Analysis of American Woodcock Nest and Brood Habitat." In D. M. Keppie and R. B. Owen, eds., *Sixth Woodcock Symposium, Held at Fredericton, New Brunswick, October 4, 5 and 6*, pp. 109-118. Laurel, Md.: Patuxent Wildlife Research Center.

Bradbury, J. W., R. M. Gibson, and I. M. Tsai. 1986. "Hotspots and the Dispersion of Leks." *Animal Behavior* 34: 1694-1709.

Brewster, W. 1894. "Notes and Song-Flight of the Woodcock (*Philohela minor*)." *Auk* (4) 11: 291-298.

Brewster, W. 1925. "Birds of the Lake Umbagog Region of Maine." *Bulletin of the Museum of Comparative Zoology at Harvard College* 66 (2). In A. C. Bent, *Life Histories of North American Shorebirds*, part 1, New York: Dover, 1962.

Bruggink, J. G., E. J. Oppelt, K. E. Doherty, D. E. Andersen, J. Meunier, and R. S. Lutz. 2013. "Fall Survival of American Woodcock in the Western Great Lakes Region." *Journal of Wildlife Management* 77 (5): 1021-1030.

Burroughs, J. 1901. *The Wit of a Duck and Other Papers*. Boston: Houghton Mifflin.

Burroughs, J. 1908. *The Writings of John Burroughs*. Vol. 15. Boston: Houghton Mifflin.

Caldwell, P. D., and J. S. Lindzey. 1974. "The Behavior of Adult Female Woodcock in Central Pennsylvania." In *Proceedings of the Fifth American Woodcock Workshop*, pp. 178–191. Athens: University of Georgia Center for Continuing Education.

Carney, T. 1993. *Sun-Drenched Days, Two-Blanket Nights*. Utica, Mich.: Partridge Pointe Press.

Carson, P. 1991. "The Future of Woodcock Hunting." In *Come October: Exclusively Woodcock*, pp. 133–153. Traverse City, Mich.: Countrysport Press.

Causey, K., G. Horton, J. Roboski, R. Johnson, and P. Mason. 1979. "American Woodcock Hatched in Alabama Killed in Michigan." *Wilson Bulletin* 91 (3): 463–464.

Chandler, C. C., D. I. King, and R. B. Chandler. 2012. "Do Mature Forest Birds Prefer Early-Successional Habitat during the Post-Fledging Period?" *Forest Ecology and Management* 264: 1–9.

Chapman, F. M. 1907. *Handbook of Birds of Eastern North America*. 7th ed. New York: D. Appleton and Co.

Clay, M. B. 1933. "Diurnal Activity of the Woodcock." *Wilson Bulletin* 45 (3): 106–110.

Cobb, S. 1959. "On the Angle of the Cerebral Axis in the American Woodcock." *Auk* 76 (1): 55–59.

Cole, L. J. 1922. "The Early History of Bird Banding in America." *Wilson Bulletin* 34 (2): 108–114.

Connors, J. I., and P. D. Doerr. 1982. "Woodcock Use of Agricultural Fields in Coastal North Carolina." In T. J. Dwyer and G. L. Storm, eds., *Woodcock Ecology and Management: Papers from the Seventh Woodcock Symposium Held at the Pennsylvania State University, University Park, 28-30 October 1980*, pp. 139–147. Washington, D.C.: U.S. Fish and Wildlife Service.

Coon, R. A., P. D. Caldwell, and G. L. Storm. 1976. "Some Characteristics of Fall Migration of Female Woodcock." *Journal of Wildlife Management* 40 (1): 91–95.

Cooper, T. R., and R. D. Rau. 2015. "American Woodcock Population Status, 2015." U.S. Fish and Wildlife Service, https://www.fws.gov/migratorybirds/pdf/surveys-and -data/Population-status/ Woodcock/AmericanWoodcockStatusReport17.pdf.

Cottam, C. 1934. "Adaptability in the Feeding Habits of the Woodcock." *Wilson Bulletin* 46 (3): 200.

Cronon, W. 1983. *Changes in the Land: Indians, Colonists, and the Ecology of New England*. New York: Hill and Wang.

Crowell, A. E. 1947. "Cape Cod Memories." In E. V. Connet, ed., *Duck Shooting along the Atlantic Tidewater*, pp. 56–60. New York: Bonanza Books.

Davis, F. W. 1970. "Territorial Conflict in the American Woodcock." *Auk* 82 (3): 327–328.

Davis, T. 2004. *The Tattered Autumn Sky*. Guilford, Conn.: Lyons Press.

Davis, W. B. 1961. "Woodcock Nesting in Brazos County, Texas." *Auk* 78 (2): 272–273.

de la Valdène, G. 1985. *Making Game: An Essay on Woodcock*. Livingston, Mont.: Clark City Press.

Derleth, E. L., and G. F. Sepik. 1990. "Summer-Fall Survival of American Woodcock in Maine." *Journal of Wildlife Management* 54 (1): 97–106.

Dessecker, D. R., and D. G. McAuley. 2001. "Importance of Early Successional Habitat to Ruffed Grouse and American Woodcock." *Wildlife Society Bulletin* 29 (2): 456–465.

Diefenbach, D. R., E. L. Derleth, W. M. V. Haegen, J. D. Nichols, and J. E. Hines. 1990. "American Woodcock Winter Distribution and Fidelity to Wintering Areas." *Auk* 107 (4): 745–749.

Dilworth, T. G., J. A. Keith, P. A. Pearce, and L. M. Reynolds. 1972. "DDE and Eggshell Thickness in New Brunswick Woodcock." *Journal of Wildlife Management* 36 (4): 1186–1193.

Dilworth, T. G., P. A. Pearce, and J. V. Dobell. 1974. "DDT in New Brunswick Woodcocks." *Journal of Wildlife Management* 38 (2): 331–337.

Dinsmore, J. J. 1994. *A Country So Full of Game: The Story of Wildlife in Iowa*. Iowa City: University of Iowa Press.

Doherty, K. E., D. E. Andersen, J. Meunier, E. Oppelt, R. S. Lutz, and J. G. Bruggink. 2010. "Foraging Location Quality as a Predictor of Fidelity to a Diurnal Site for Adult Female American Woodcock, *Scolopax minor*." *Wildlife Society Bulletin* 16 (4): 379–388.

Domke, G. M., A. R. Ek, M. A. Kilgore, and A. J. David. 2008. "Aspen in the Lakes States: A Research Review." *National Council for Air and Stream Improvement Technical Bulletin* 955.

Dorsey, C. 1990. *The Grouse Hunter's Almanac*. Stillwater, Minn.: Voyageur Press.

Douglas, M. R., J. R. Rohr, and J. F. Tooker. 2015. "Neonicotinoid Insecticide Travels through a Soil Food Chain, Disrupting Biological Control of Non-Target Pests and Decreasing Soya Bean Yield." *Journal of Applied Ecology* 52: 250–260.

Duke, G. E. 1966. "Reliability of Censuses of Singing Male Woodcock." *Journal of Wildlife Management* 30 (4): 697–707.

Dunford, R. D., and R. B. Owen. 1973. "Summer Behavior of Immature Radio-Equipped Woodcock in Central Maine." *Journal of Wildlife Management* 37 (4): 462–469.

Dwyer, T. J., E. L. Derleth, and D. G. McAuley. 1982. "Woodcock Brood Ecology in Maine." In T. J. Dwyer and G. L. Storm, eds., *Woodcock Ecology and Management:*

Papers from the Seventh Woodcock Symposium Held at the Pennsylvania State University, University Park, 28-30 October 1980, pp. 55-62. Washington, D.C.: U.S. Fish and Wildlife Service.

Dwyer, T. J., D. G. McAuley, and E. L. Derleth. 1983. "Woodcock Singing-Ground Counts and Habitat Changes in the Northeastern United States." *Journal of Wildlife Management* 47 (3): 772-779.

Dwyer, T. J., and J. D. Nichols. 1982. "Regional Population Inferences for the American Woodcock." In T. J. Dwyer and G. L. Storm, eds., *Woodcock Ecology and Management: Papers from the Seventh Woodcock Symposium Held at the Pennsylvania State University, University Park, 28-30 October 1980*, pp. 12-21. Washington, D.C.: U.S. Fish and Wildlife Service.

Dwyer, T. J., G. F. Sepik, E. L. Derleth, and D. G. McAuley. 1988. "Demographic Characteristics of a Maine Woodcock Population and Effects of Habitat Management." Washington, D.C.: U.S. Fish and Wildlife Service.

Dyer, J. M., and R. B. Hamilton. 1974. "An Analysis of Feeding Habits of the American Woodcock (*Philohela minor*) in Southern Louisiana." In *Proceedings of the Fifth American Woodcock Workshop*, pp. 286-296. Athens: University of Georgia Center for Continuing Education.

Earnest, A. 1982. *The Art of the Decoy: American Bird Carving*. Exton, Penn.: Schiffer Publishing.

Edminster, F. C. 1954. *American Game Birds of Field and Forest: Their Habits, Ecology, and Management*. New York: Castle Books.

Evans, G. B. 1971. *The Upland Shooting Life*. New York: Alfred A. Knopf.

Fergus, C. 1992. *A Hunter's Road*. New York: Henry Holt and Co.

Fergus, C. 2005. *A Hunter's Book of Days*. Camden, Maine: Countrysport Press.

Ferling, J. E. 1992. *John Adams: A Life*. New York: Oxford University Press.

Fisher, A. K. 1901. "Two Vanishing Game Birds—the Woodcock and the Wood Duck." In *Yearbook of the United States Department of Agriculture for 1901*, pp. 447-458. Washington, D.C.: G.P.O.

Flanigan, T. 2013. "Sharing the Experience." *Ruffed Grouse Society* 25 (Spring): 18, 20.

Forbush, E. H. 1916. *A History of the Game Birds, Wild-Fowl, and Shore Birds of Massachusetts and Adjacent States*. Boston: Massachusetts State Board of Agriculture.

Ford, C. 1996. *The Trickiest Thing in Feathers*. L. Morrow, ed. Gallatin, Mont.: Wilderness Adventures Press.

Forester, F. 1951. *On Upland Shooting*. A. R. Beverley-Giddings, ed. New York: William Morrow.

Freeman, M. 2013. "In Theory and in Practice: What Makes a Good Clearcut?" *Northern Woodlands* 20 (Winter): 36–45.

Frelich, L. E., C. M. Hale, S. Scheu, A. R. Holdsworth, L. Heneghan, P. J. Bohlen, and P. B. Reich. 2006. "Earthworm Invasion into Previously Earthworm-Free Temperate and Boreal Forests." *Biological Invasions* 8: 1235–1245.

Glasgow, L. L. 1958. "Contributions to the Knowledge of the Ecology of the American Woodcock, *Philohela minor*, on the Wintering Range in Louisiana." PhD dissertation, Texas A&M University.

Gobster, P. H. 2001. "Human Dimensions of Early Successional Landscapes in the Eastern United States." *Wildlife Society Bulletin* 29 (2): 474–482.

Gosler, A. G., J. P. Higham, and S. J. Reynolds. 2005. "Why Are Birds' Eggs Speckled?" *Ecology Letters* 8: 1105–1113.

Gould, S. J. 1980. *The Panda's Thumb: More Reflections in Natural History*. New York: W. W. Norton.

Goulson, D. 2013. "An Overview of the Environmental Risks Posed by Neonicotinoid Pesticides." *Journal of Applied Ecology* 50: 977–987.

Grant, M. C. 1993. "The Trembling Giant." *Discover* (October): 83–88.

Greeley, F. 1953. "Sex and Age Studies of Fall-Shot Woodcock (*Philohela minor*) from Southern Wisconsin." *Journal of Wildlife Management* 17 (1): 29–32.

Gregg, L. E. 1982. "Woodcock Singing Ground Counts and Breeding Habitat." In T. J. Dwyer and G. L. Storm, eds., *Woodcock Ecology and Management: Papers from the Seventh Woodcock Symposium Held at the Pennsylvania State University, University Park, 28–30 October 1980*, pp. 30–33. Washington, D.C.: U.S. Fish and Wildlife Service.

Gregg, L. E., and J. B. Hale. 1977. "Woodcock Nesting Habitat in Northern Wisconsin." *Auk* 94 (3): 489–493.

Gregory, J. F., and R. M. Whiting, Jr. 2000. "Food Habits and Preference of American Woodcock in East Texas Pine Plantations." In D. G. McAuley, J. G. Bruggink, and G. F. Sepik, eds., *Proceedings of the Ninth American Woodcock Symposium*, pp. 23–35. Laurel, Md.: Patuxent Wildlife Research Center.

Gregory, S. S., Jr. n.d. Stephen S. Gregory file, Chicago Academy of Sciences. In J. Greenberg, ed., *Of Prairie, Woods, and Water: Two Centuries of Chicago Nature Writing*, pp. 233–234. Chicago: University of Chicago Press, 2008.

Grinnell, G. B. 1910. *American Game-Bird Shooting*. New York: Forest and Stream.

Grinnell, G. B. 1922. "Woodcock Carrying Its Young." *Auk* 39 (4): 563–564.

Gutzwiller, K. J., K. R. Kinsley, G. L. Storm, W. M. Tzilkowski, and J. S. Wakeley. 1983.

"Relative Value of Vegetation Structure and Species Composition for Identifying American Woodcock Breeding Habitat." *Journal of Wildlife Management* 47 (2): 535-540.

Gutzwiller, K. J., and J. S. Wakeley. 1982. "Differential Use of Woodcock Singing Grounds in Relation to Habitat Characteristics." In T. J. Dwyer and G. L. Storm, eds., *Woodcock Ecology and Management: Papers from the Seventh Woodcock Symposium Held at the Pennsylvania State University, University Park, 28-30 October 1980*, pp. 51-54. Washington, D.C.: U.S. Fish and Wildlife Service.

Hagan, J. M. 1993. "Decline of the Rufous-Sided Towhee in the Eastern United States." *Auk* 110 (4): 863-874.

Hale, C. M., L. E. Frelich, and P. B. Reich. 2005. "Exotic European Earthworm Invasion Dynamics in Northern Hardwood Forests in Minnesota, USA." *Ecological Applications* 15: 848-860.

Hale, J. B., and L. E. Gregg. 1976. "Woodcock Use of Clearcut Aspen Areas in Wisconsin." *Wildlife Society Bulletin* 4 (3): 111-115.

Hamilton, W. D., and M. Zuk. 1982. "Heritable True Fitness and Bright Birds: A Role for Parasites." *Science* 218: 384-387.

Harrington, B. A. 1999. "The Hemispheric Globetrotting of the White-Rumped Sandpiper." In K. A. Able, ed., *Gatherings of Angels: Migrating Birds and Their Ecology*, pp. 119-134. Ithaca, N.Y.: Cornell University Press.

Harris, D., C. Elliott, and R. Frederick. 2009. "Habitat Characteristics Associated with American Woodcock (*Scolopax minor* Gmelin) Nests in Central Kentucky." *Journal of the Kentucky Academy of Sciences* 70 (2): 141-144.

Haukos, D. A., R. M. Whiting, Jr., and L. M. Smith. 2007. "Population Characteristics of American Woodcock Wintering in Texas." *Proceedings of the Annual Conference of Southeast Association of Fish and Wildlife Agencies* 61: 82-88.

Hedenström, A. 2010. "Extreme Endurance Migration: What Is the Limit of Non-Stop Flight?" *PLOS Biology* 8: 1-6.

Heilner, V. C. 1941. *Our American Game Birds*. Garden City, N.Y.: Doubleday.

Heinrich, B. 2016. "Notes on the Woodcock Rocking Display." *Northeastern Naturalist* 23 (1): N4-N7.

Henderson, C. L. 2008. *Birds in Flight: The Art and Science of How Birds Fly*. Minneapolis: Voyageur Press.

Herwig, M. 2016. "Woodcock Way: Find Woodcock Near the Jack Pines." *Skydance* 4 (1): 24-25.

Hill, G., and S. Smith. 1981. *The Whispering Wings of Autumn*. Bozeman, Mont.: Wilderness Adventures Press.

Hodgson, B. 1994. "Buffalo: Back Home on the Range." *National Geographic* 185 (5): 64–89.

Holdsworth, A. R., L. E. Frelich, and P. B. Reich. 2007. "Effects of Earthworm Invasion on Plant Species Richness in Northern Hardwood Forests." *Conservation Biology* 21 (4): 997–1008.

Holland, D. 1961. *The Upland Game Hunter's Bible*. Garden City, N.Y.: Doubleday.

Holland, R. P. 1944. *Shotgunning in the Uplands*. New York: Countryman Press.

Hornaday, W. T. 1904. *Hornaday's American Natural History*. New York: Charles Scribner's Sons.

Horton, G. I., and M. K. Causey. 1979. "Woodcock Movements and Habitat Utilization in Central Alabama." *Journal of Wildlife Management* 43 (2): 414–420.

Horton, G. I., and M. K. Causey. 1981. "Dispersal of American Woodcock in Central Alabama after Brood Breakup." *Journal of Wildlife Management* 45 (4): 1058–1061.

Horton, G. I., and M. K. Causey. 1984. "Brood Abandonment by Radio-Tagged American Woodcock Hens." *Journal of Wildlife Management* 48 (2): 606–607.

Hudgins, J. E., G. L. Storm, and J. S. Wakeley. 1985. "Local Movements and Diurnal-Habitation Selection by Male American Woodcock in Pennsylvania." *Journal of Wildlife Management* 49 (3): 614–619.

Huggler, T. 1996. *A Fall of Woodcock*. Traverse City, Mich.: Countrysport Press.

Huggler, T. 2017. "Luck and the Sky Dance." *Shooting Sportsman*, March–April: 104.

Hunt, P. D. 1998. "Evidence of a Landscape Population Model of the Importance of Early Successional Habitat to the American Redstart." *Conservation Biology* 12 (6): 1377–1389.

Hunter, W. C., D. A. Buehler, R. A. Canterbury, J. L. Confer, and P. B. Hamel. 2001. "Conservation of Disturbance-Dependent Birds in Eastern North America." *Wildlife Society Bulletin* 29 (2): 440–455.

Huntington, D. W. 1903. *Our Feathered Game: A Handbook of the North American Game Birds*. London: Bickers and Son.

Jackson, J. A. 2007. *George Miksch Sutton: Artist, Scientist, and Teacher*. Norman: University of Oklahoma Press.

Jarvis, W. 1890. "The Woodcock." In W. B. Leffingwell, ed., *Shooting on Upland, Marsh, and Stream*, pp. 31–60. New York: Rand, McNally.

Johnson, D. R. 1984. "American Woodcock Carrying Young." *Loon* 56 (Spring): 66–67.

Johnson, R. C., and M. K. Causey. 1982. "Use of Longleaf Pine Stands by Woodcock in

Southern Alabama Following Prescribed Burning." In T. J. Dwyer and G. L. Storm, eds., *Woodcock Ecology and Management: Papers from the Seventh Woodcock Symposium Held at the Pennsylvania State University, University Park, 28–30 October 1980*, pp. 120–125. Washington, D.C.: U.S. Fish and Wildlife Service.

Jones, R. F. 1996. *Dancers in the Sunset Sky: The Musings of a Bird Hunter*. New York: Lyons and Burford.

Jones, R. F. 2002. *The Hunter in My Heart: A Sportsman's Salmagundi*. Guilford, Conn.: Lyons Press.

Judd, E. T. 1937. "The Woodcock in North Dakota." *Wilson Bulletin* 49 (2): 119–120

Keer, T. 2016. "Why We Hunt Woodcock." *Ruffed Grouse Society* 28 (4): 17–19.

Keppie, D. M., and G. W. Redmond. 1985. "Body Weight and Possession of Territory for Male American Woodcock." *Condor* 87 (2): 287–290.

Keppie, D. M., W. R. Watt, and G. W. Redmond. 1984. "Male Woodcock in Coniferous Forests: Implications for Route Allocations in Survey." *Wildlife Society Bulletin* 12 (2): 174–178.

Kerlinger, P. 1995. *How Birds Migrate*. Mechanicsburg, Penn.: Stackpole Books.

Kilner, R. M. 2006. "The Evolution of Egg Colour and Patterning in Birds." *Biological Reviews* 81: 383–406.

Kimball, J. W. 1970. *The Spirit of the Wilderness*. Minneapolis: T. S. Denison and Co.

Kletzly, R. C. 1976. "American Woodcock in West Virginia." Charleston: West Virginia Department of Natural Resources.

Knight, J. A. 1946. *Woodcock*. New York: Alfred A. Knopf.

Krementz, D. G., and J. B. Berdeen. 1997. "Survival Rates of American Woodcock Wintering in the Georgia Piedmont." *Journal of Wildlife Management* 61 (4): 1328–1332.

Krementz, D. G., R. Crossett, and S. E. Lehnen. 2014. "Nocturnal Field Use by Fall Migrating American Woodcock in the Delta of Arkansas." *Journal of Wildlife Management* 78 (2): 264–272.

Krementz, D. G., J. E. Hines, and D. R. Luukkonen. 2003. "Survival and Recovery Rates of American Woodcock Banded in Michigan." *Journal of Wildlife Management* 67 (2): 398–405.

Krementz, D. G., and J. J Jackson. 1999. "Woodcock in the Southeast: Natural History and Management for Landowners." Athens: University of Georgia College of Agricultural and Environmental Sciences Cooperative Extension Service.

Krementz, D. G., J. T. Seginak, and G. W. Pendleton. 1994. "Winter Movements and Spring Migration of American Woodcock along the Atlantic Coast." *Wilson Bulletin* 106 (3): 482–493.

Krementz, D. G., J. T. Seginak, and G. W. Pendleton. 1995. "Habitat Use at Night by Wintering American Woodcock in Coastal Georgia and Virginia." *Wilson Bulletin* 107 (4): 686–697.

Krementz, D. G., J. T. Seginak, D. R. Smith, and G. W. Pendleton. 1994. "Survival Rates of American Woodcock Wintering along the Atlantic Coast." *Journal of Wildlife Management* 58 (1): 147–155.

Krohn, W. B. 1971. "Some Patterns of Woodcock Activities on Maine Summer Fields." *Wilson Bulletin* 83 (4): 396–407.

Krohn, W. B., and E. R. Clark. 1977. "Band-Recovery Distribution of Eastern Maine Woodcock." *Wildlife Society Bulletin* 5 (3): 118–122.

Krohn, W. B., J. C. Rieffenberger, and F. Ferrigno. 1977. "Fall Migration of Woodcock at Cape May, New Jersey." *Journal of Wildlife Management* 41 (1): 104–111.

Kroodsma, D. 2005. *The Singing Life of Birds: The Art and Science of Listening to Birds.* Boston: Houghton Mifflin.

Leopold, A. 1933. *Game Management.* New York: Charles Scribner's Sons.

Leopold, A. 1949. *A Sand County Almanac.* New York: Oxford University Press.

Leopold, A. 1999. *For the Health of the Land: Previously Unpublished Essays and Other Writings.* J. B. Callicott and E. T. Freyfogle, eds. Washington, D.C.: Island Press.

Leopold, A. S., S. A. Cain, C. M. Cottam, I. N. Gabrielson, and T. L. Kimball. 1963. "Wildlife Management in the National Parks." Washington, D.C.: National Park Service.

Leopold, E. A. 2016. *Stories from the Leopold Shack: Sand County Revisited.* New York: Oxford University Press.

Levinson, J. M., and S. G. Headley. 1991. *Shorebirds: The Birds, the Hunters, the Decoys.* Centreville, Md.: Tidewater Publishers.

Lewis, E. J. 1906. *The American Sportsman.* Philadelphia: Lippincott and Co.

Lincoln, F. C. 1921. "The History and Purpose of Bird Banding." *Auk* 38 (2): 217–228.

Lipske, M. 1997. "Forests Rise Woodcocks Fall: The Return of Mature Forests in the East, Good News for Most Wildlife, May Be Bad News for This Scrub-Loving Bird." Washington, D.C.: National Wildlife Federation.

Liscinsky, S. A. 1965. *The American Woodcock in Pennsylvania.* Harrisburg: Pennsylvania Game Commission.

Liscinsky, S. A. 1972. *The Pennsylvania Woodcock Management Study.* Harrisburg: Pennsylvania Game Commission.

Liscinsky, S. A., and W. J. Bailey, Jr. 1955. "A Modified Shorebird Trap for Capturing Woodcock and Grouse." *Journal of Wildlife Management* 19 (3): 405–408.

Litvaitis, J. A. 1993. "Response of Early Successional Vertebrates to Historic Changes in Land Use." *Conservation Biology* 7 (4): 866–873.

Long, A. K., and A. Locher. 2013. "American Woodcock (*Scolopax minor*) Use of Pine Plantation Habitat during Spring in Central Arkansas." *Wilson Journal of Ornithology* 125 (2): 322–328.

Long, W. J. 1903. *A Little Brother to the Bear*. In R. H. Lutts, *The Nature Fakers: Wildlife, Science, and Sentiment*, p. 76. Golden, Colo.: Fulcrum, 1990.

Lorimer, C. G. 2001. "Historical and Ecological Roles of Disturbance in Eastern North American Forests: 9,000 Years of Change." *Wildlife Society Bulletin* 29 (2): 425–439.

Loucks, O. L. 1983. "New Light on the Changing Forest." In S. L. Flader, ed., *The Great Lakes Forest: An Environmental and Social History*, pp. 17–33. Minneapolis: University of Minnesota Press.

Lundrigan, T. N. 2006. *A Bird in the Hand*. Camden, Maine: Countrysport Press.

Mabee, T. J., A. M. Wildman, and C. B. Johnson. 2006. "Using Egg Flotation and Eggshell Evidence to Determine Age and Fate of Arctic Shorebird Nests." *Journal of Field Ornithology* 77 (2): 163–172.

Mackey, W. F. 1965. *American Bird Decoys*. New York: E. P Dutton.

MacQuarrie, G. 1940. "Old Deacon Woodcock." In Ellen Gibson-Wilson, ed., *The Gordon MacQuarrie Sporting Treasury*. Minocqua, Wis.: Willow Creek Press, 1998.

Marsh, George Perkins. 1864. *Man and Nature: Or, Physical Geography as Modified by Human Action*. New York: Charles Scribner's Sons.

Marshall, W. H. 1982a. "Does the Woodcock Bob or Rock—and Why?" *Auk* 99 (4): 791–792.

Marshall, W. H. 1982b. "Minnesota Woodcock." *Loon* 54 (Winter): 203–211.

Martin, F. W. 1964. "Woodcock Age and Sex Determination from Wings." *Journal of Wildlife Management* 28 (2): 287–293.

Masotti, P. 2015. "A Cajun Connection." *Ruffed Grouse Society* 27 (4): 19–23.

Masse, R. J., B. C. Tefft, and S. R. McWilliams. 2014. "Multiscale Habitat Selection by a Forest Dwelling Shorebird, the American Woodcock: Implications for Forest Management in Southern New England, USA." *Forest Ecology and Management* 325: 37–48.

Masse, R. J., B. C. Tefft, and S. R. McWilliams. 2015. "Higher Bird Abundance and Diversity Where American Woodcock Sing: Fringe Benefits of Managing Forests for Woodcock." *Journal of Wildlife Management* 79 (8): 1378–1384.

Mathewson, W. 2000. *Best Birds Upland and Shore*. Amity, Ore.: Sand Lake Press.

Matthiessen, P. 1967. *The Wind Birds: Shorebirds of North America*. Shelburne, Vt.: Chapters.

McAuley, D. G., J. R. Longcore, and G. F. Sepik. 1990. "Renesting by American Wood-cocks (*Scolopax minor*) in Maine." *Auk* 107 (2): 407–410.

McAuley, D. G., J. R. Longcore, G. F. Sepik, and G. W. Pendleton. 1996. "Habitat Selection of American Woodcock Nest Sites on a Managed Area in Maine." *Journal of Wildlife Management* 60 (1): 138–148.

McCabe, R. A. 1982. "The American Woodcock: A Keynote Address." In T. J. Dwyer and G. L. Storm, eds., *Woodcock Ecology and Management: Papers from the Seventh Woodcock Symposium Held at the Pennsylvania State University, University Park, 28–30 October 1980*, pp. 1–6. Washington, D.C.: U.S. Fish and Wildlife Service.

McCabe, R. A. 1987. *Aldo Leopold, the Professor*. Amherst, Wis.: Palmer Publications.

McDermid, C. C. 1914. "The Woodcock Carrying Its Young." *Auk* 31 (3): 398–399.

McFarlane, R. W. 1992. *A Stillness in the Pines: The Ecology of the Red-Cockaded Woodpecker*. New York: W. W. Norton.

McHenry, A. G. 1983. *Becasse, the American Woodcock in Louisiana*. Washington, D.C.: Published in agreement with the U.S. Fish and Wildlife Service.

McIntosh, M. 1996. *Wild Things*. New Albany, Ohio: Countrysport Press.

McIntosh, M. 1997. *Traveler's Tales: The Wanderings of a Bird Hunter and Sometime Fly Fisherman*. Camden, Maine: Down East Books.

McLeish, T. 2007. *Golden Wings and Hairy Toes: Encounters with New England's Most Imperiled Wildlife*. Lebanon, N.H.: University Press of New England.

Mendall, H. L. 1938. "A Technique for Banding Woodcock." *Bird-Banding* 9 (3): 153–155.

Mendall, H. L., and C. M. Aldous. 1943. *The Ecology and Management of the American Woodcock*. Orono: Maine Cooperative Wildlife Research Unit.

Merritt, C. 1904. *The Shadow of a Gun*. Chicago: F. T. Peterson Co.

Meunier, J., R. Song, R. S. Lutz, D. E. Andersen, K. E. Doherty, J. G. Bruggink, and E. Oppelt. 2008. "Proximate Cues for a Short-Distance Migratory Species: An Application of Survival Analysis." *Journal of Wildlife Management* 72 (2): 440–448.

Miller, D. L., and M. K. Causey. 1985. "Food Preferences of American Woodcock Wintering in Alabama." *Journal of Wildlife Management* 49 (2): 492–496.

Miller, H. D., and M. J. Jordan. 2011. "Relationship between Exotic Invasive Shrubs and American Woodcock (*Scolopax minor*) Nest Success and Habitat Selection." *Journal of the Pennsylvania Academy of Science* 85 (4): 132–139.

Mitten, J. B., and M. C. Grant. 1996. "Genetic Variation and the Natural History of Quaking Aspen." *Bioscience* 46 (1): 25–31.

Moore, J. 2016. "Following the American Woodcock: How Satellite Transmitters Help Improve Habitat Management." *Skydance* 4 (1): 8–9.

Muir, J. 1911. *My First Summer in the Sierra.* New York: Houghton Mifflin.

Muir, J. 1913. *The Story of My Boyhood and Youth.* New York: Houghton Mifflin..

Mulak, S. J. 1987. *Brown Feathers: Waterfowling Tales and Upland Dreams.* Harrisburg, Penn.: Stackpole Books.

Murphy, D. W., and F. R. Thompson. 1993. "Breeding Chronology of the American Woodcock in Missouri." In J. R. Longore and G. F. Sepik, eds., *Eighth Woodcock Symposium,* pp. 12–18. Washington, D.C.: U.S. Fish and Wildlife Service.

Myatt, N. A., and D. G. Krementz. 2007a. "American Woodcock Fall Migration Using Central Region Band-Recovery and Wing-Collection Survey Data." *Journal of Wildlife Management* 71 (2): 336–343.

Myatt, N. A., and D. G. Krementz. 2007b. "Fall Migration and Habitat Use of American Woodcock in the Central United States." *Journal of Wildlife Management* 71 (4): 1197–1205.

Nelson, M. R., and D. E. Andersen. 2013. "Do Singing-Ground Surveys Reflect American Woodcock Abundance in the Western Great Lakes Region?" *Wildlife Society Bulletin* 37 (3): 585–595.

Norris, R. T., J. D. Beule, and A. T. Studholme. 1940. "Banding Woodcock on Pennsylvania Singing Grounds." *Journal of Wildlife Management* 4 (1): 8–14.

Olmstead, R. O. 1951. "A Spring Record of the Woodcock, *Philohela minor,* in Kansas." *Auk* 68 (3): 375.

Osborn, J. 2016. "Autumn in the Wings." *Upland Almanac,* Spring: 43.

Owen, R. B., J. M. Anderson, J. W. Artmann, E. R. Clark, T. G. Dilworth, L. E. Gregg, F. W. Martin, J. D. Newsom, and S. R. Pursglove. 1977. "American Woodcock." In G. C. Sanderson, ed., *Management of Migratory Shore and Upland Game Birds in North America,* pp. 149–188. Washington, D.C.: International Association of Fish and Wildlife Agencies.

Owen, R. B., and W. J. Galbraith. 1989. "Earthworm Biomass in Relation to Forest Type, Soil, and Land Use: Implications for Woodcock Management." *Wildlife Society Bulletin* 17 (2): 130–136.

Owen, R. B., and J. W. Morgan. 1975a. "Influence of Night-Lighting and Banding on Woodcock Movements." *Wildlife Society Bulletin* 3 (2): 77–79.

Owen, R. B., and J. W. Morgan. 1975b. "Summer Behavior of Adult Radio-Equipped Woodcock in Central Maine." *Journal of Wildlife Management* 39 (1): 179–182.

Pace, R. M. 2000. "Winter Survival Rates of American Woodcock in South Central Louisiana." *Journal of Wildlife Management* 64 (4): 933-939.

Pan, Y., J. M. Chen, R. Birdsey, K. McCullough, L. He, and F. Deng. 2011. "Age Structure and Disturbance Legacy of North American Forests." *Biogeosciences* 8: 715-732.

Parman, M. 2015. "The Looney Bird." *Ruffed Grouse Society* 27 (Fall): 43-45.

Peek, J. M., D. L. Ulrich, and R. J. Mackie. 1976. "Moose Habitat Selection and Relationships to Forest Management in Northeastern Minnesota." *Wildlife Monograph* 48 (April): 3-65.

Peterson, R. T. 2006. *All Things Reconsidered: My Birding Adventures.* B. Thompson, ed. Boston: Houghton Mifflin.

Peterson, R. T., and J. Fisher. 1955. *Wild America: The Legendary Story of Two Great Naturalists on the Road.* Boston: Houghton Mifflin.

Pettingill, O. S. 1936. "The American Woodcock." *Memoirs of the Boston Society of Natural History* 9: 169-391.

Pettingill, O. S. 1939. "Additional Information on the Food of the American Woodcock." *Wilson Bulletin* 51 (2): 78-82.

Pitelka, F. A. 1943. "Territoriality, Display, and Certain Ecological Relations of the American Woodcock." *Wilson Bulletin* 55 (2): 88-114.

Porneluzi, P. A., R. Brito-Aguilar, R. L. Clawson, and J. Faaborg. 2014. "Long-Term Dynamics of Bird Use in Clearcuts in Post-Fledging Period." *Wilson Journal of Ornithology* 126 (4): 623-634.

Primack, P. 2016. "Four Decades of Management." *Northern Woodlands* 23 (Spring): 20-21.

Pyne, S. 1982. *Fire in America: A Cultural History of Wildland and Rural Fire.* Princeton, N.J.: Princeton University Press.

Rabe, D. L. 1977. "Structural Analysis of Woodcock Diurnal Habitat in Northern Michigan." In D. M. Keppie and R. B. Owen, eds., *Sixth Woodcock Symposium, Held at Fredericton, New Brunswick, October 4, 5 and 6*, pp. 125-134. Laurel, Md.: Patuxent Wildlife Research Center.

Rabe, D. L., and H. H. Prince. 1982. "Breeding Woodcock Use of Manipulated Forest-Field Complexes in Aspen Communities." In T. J. Dwyer and G. L. Storm, eds., *Woodcock Ecology and Management: Papers from the Seventh Woodcock Symposium Held at the Pennsylvania State University, University Park, 28-30 October 1980*, pp. 114-119. Washington, D.C.: U.S. Fish and Wildlife Service.

Rabe, D. L., H. H. Prince, and D. L. Beaver. 1983. "Feeding-Site Selection and Foraging Strategies of American Woodcock." *Auk* 100 (3): 711-716.

Rabe, D. L., H. H. Prince, and E. D. Goodman. 1983. "The Effects of Weather on Bioenergetics of Breeding American Woodcock." *Journal of Wildlife Management* 47 (3): 762-771.

Rapai, W. 2012. *The Kirtland's Warbler: The Story of a Bird's Fight against Extinction and the People Who Saved It*. Ann Arbor: University of Michigan Press.

Rhymer, J. M., D. G. McAuley, H. L. Ziel. 2005. "Phylogeography of the American Woodcock (*Scolopax minor*): Are Management Units Based on Band Recovery Data Reflected in Genetically Based Management Units?" *Auk* 122 (4): 1149-1160.

Ripley, O. 1926. *Sport in Field and Forest*. New York: D. Appleton and Co.

Roberts, T. H., E. P. Hill, and E. A. Gluesing. 1984. "Woodcock Utilization of Bottom-land Hardwoods in the Mississippi Delta." *Proceedings of the Annual Conference of Southeast Association of Fish and Wildlife Agencies* 38: 137-141.

Roboski, J. C., and M. K. Causey. 1981. "Incidence, Habitat Use, and Chronology of Woodcock Nesting in Alabama." *Journal of Wildlife Management* 45 (3): 793-797.

Rodewald, P. G., and M. C. Brittingham. 2004. "Stopover Habitats of Landbirds during the Fall: Use of Edge Dominated and Early-Successional Forests." *Auk* 121 (4): 1040-1055.

Rosene, W. 1949. "Woodcock at Sea." *Wilson Bulletin* 61 (4): 235-236.

Rutledge, A. 1935. *Wild Life of the South*. In J. Casada, ed., *Hunting and Home in the Southern Heartland: The Best of Archibald Rutledge*, p. 100. Columbia: University of South Carolina Press, 1992.

Samuel, D. E., and D. R. Beightol. 1973. "The Vocal Repertoire of Male American Wood-cock." *Auk* 90 (4): 906-909.

Sandys, E. W. 1890. "Woodcock Shooting in Canada." In W. Mathewson, ed., *Upland Tales*, pp. 1-8. Long Beach, Calif.: Safari Press, 1992.

Sandys, E. W. 1899. "The Woodcock and His Ways." *Outings* 37 (1): 19-23.

Sandys, E. W., and T. S. Van Dyke. 1924. *Upland Game Birds*. New York: Macmillan.

Sauer, J. R., and J. B. Bortner. 1991. "Population Trends from the American Woodcock Singing-Ground Survey, 1970-1988." *Journal of Wildlife Management* 55 (2): 300-312.

Scheuhammer, A. M., C. A. Rogers, and D. E. Bond. 1999. "Elevated Lead Exposure in American Woodcock in Eastern Canada." *Archives of Environmental Contaminants and Toxicology* 36: 334-340.

Scheuhammer, A. M., D. E. Bond, N. M. Burgess, and J. Rodriguez. 2003. "Lead and Stable Lead Isotope Ratios in Soils, Earthworms, and Bones of American Woodcock (*Scolopax minor*) in Eastern Canada." *Environmental Toxicology and Chemistry* 22: 2585-2591.

Schlossberg, S., and D. I. King. 2009. "Postlogging Succession and Habitat Usage by Shrubland Birds." *Journal of Wildlife Management* 73 (2): 226-231.

Schorger, A. W. 1929. "Woodcock Carrying Young." *Auk* 46 (2): 232.

Seamans, M. E., and R. D. Rau. 2017. "American Woodcock Population Status, 2017." U.S. Fish and Wildlife Service, https://www.fws.gov/migratorybirds/pdf/surveys -and-data/Population-status/Woodcock/AmericanWoodcockStatusReport17.pdf.

Sepik, G. F. 1994. "A Woodcock in the Hand." Coraopolis, Penn.: Ruffed Grouse Society.

Sepik, G. F. and E. L. Derleth. 1993. "Premigratory Dispersal and Fall Migration of American Woodcock in Maine." In J. R. Longcore and G. F. Sepik, eds., *Eighth Woodcock Symposium*, pp. 36–40. Washington, D.C.: U.S. Fish and Wildlife Service.

Sepik, G. F., and T. J. Dwyer. 1982. "Woodcock Response to Habitat Management in Maine." In T. J. Dwyer and G. L. Storm, eds., *Woodcock Ecology and Management: Papers from the Seventh Woodcock Symposium Held at the Pennsylvania State University, University Park, 28–30 October 1980*, pp. 106–113. Washington, D.C.: U.S. Fish and Wildlife Service.

Sepik, G. F., R. B. Owen, and M. W. Coulter. 1983. "A Landowner's Guide to Woodcock Management in the Northeast—Adopted for Wisconsin." Moosehorn National Wildlife Refuge, U.S. Fish and Wildlife Service Miscellaneous Publication 253.

Sharp, D. 2015. "Landowners Managing Habitat to Help Canada Lynx in Maine." *Great Falls Tribune*, September 7.

Sheldon, H. P. 1993. "Woodcock Shooting." In E. V. Connett, ed., *Upland Game Bird Shooting in America*, pp. 43–58. Lyon, Miss.: Derrydale Press.

Sheldon, W. G. 1955. "Methods of Trapping Woodcocks on Their Breeding Grounds." *Journal of Wildlife Management* 19 (1): 109–115.

Sheldon, W. G. 1961. "Summer Crepuscular Flights of American Woodcock in Central Massachusetts." *Wilson Bulletin* 73 (2): 126–139.

Sheldon, W. G. 1967. *The Book of the American Woodcock*. Amherst: University of Massachusetts Press.

Sheldon, W. G., F. Greeley, and J. Krupa. 1958. "Aging Fall-Shot American Woodcocks by Primary Wear." *Journal of Wildlife Management* 22 (3): 310–312.

Shissler, B. P., and D. E. Samuel. 1983. "Observations of Male Woodcock on Singing Grounds." *Wilson Bulletin* 95 (4): 655–656.

Shissler, B. P., and D. E. Samuel. 1985. "Effectiveness of American Woodcock Survey Routes in Detecting Active Singing Grounds." *Wildlife Society Bulletin* 13 (2): 157–160.

Silvestro, R. 2007. "Cat in a Quandary." *National Wildlife Federation Magazine* 46 (1): 24–30.

Sisley, N. 1980. *Grouse and Woodcock: An Upland Hunter's Book.* Harrisburg, Penn.: Stackpole Books.

Smith, R. D., and J. S. Barclay. 1978. "Evidence of Westward Changes in the Range of the American Woodcock." *American Birds* 32 (6): 1122–1127.

Smith, R. J., and M. I. Hatch. 2008. "A Comparison of Shrub-Dominated and Forested Habitat Use by Spring Migrating Landbirds in Northeastern Pennsylvania." *Condor* 110 (4): 682–693.

Spiller, B. 1935. *Grouse Feathers.* Lanham, Md.: Derrydale Press.

Spiller, B. 1972. *More Grouse Feathers.* New York: Crown.

Stewart, O. 2002. *Forgotten Fires: Native Americans and the Transient Wilderness.* Norman: University of Oklahoma Press.

Stickel, W. H., W. E. Dodge, W. G. Sheldon, J. B. DeWitt, and L. F. Stickel. 1965. "Body Condition and Response to Pesticides in Woodcocks." *Journal of Wildlife Management* 29 (1): 147–155.

Stickel, W. H., D. W. Hayne, and L. F. Stickel. 1965. "Effects of Heptachlor-Contaminated Earthworms on Woodcock." *Journal of Wildlife Management* 29 (1): 132–146.

Straw, J. A., J. S. Wakeley, and J. E. Hudgins. 1986. "A Model of Management of Diurnal Habitat for American Woodcock in Pennsylvania." *Journal of Wildlife Management* 50 (3): 378–383.

Streby, H. M., and D. E. Andersen. 2012. "Movement and Cover-Type Selection by Fledgling Ovenbirds (*Seiurus aurocapilla*) after Independence from Adult Care." *Wilson Journal of Ornithology* 124 (3): 620–625.

Streby, H. M., S. M. Peterson, T. L. McAllister, and D. E. Andersen. 2011. "Use of Early-Successional Managed Northern Forest by Mature-Forest Species during the Post-Fledging Period." *Condor* 113 (4): 817–824.

Stribling, H. L., and P. D. Doerr. 1985a. "Characteristics of American Woodcock Wintering in Eastern North Carolina." *North American Bird Bander* 10 (3): 68–72.

Stribling, H. L., and P. D. Doerr. 1985b. "Nocturnal Use of Fields by American Woodcock." *Journal of Wildlife Management* 49 (2): 485–491.

Sullins, D. S., and W. C. Conway. 2013. "Winter in the Pineywoods." *Skydance* 1: 2–10.

Sullins, D. S., W. C. Warren, D. A. Haukos, K. A. Hobson, L. I. Wassenaar, C. E. Comer, and I. Hung. 2016. "American Woodcock Migratory Connectivity as Indicated by Hydrogen Isotopes." *Journal of Wildlife Management* 80 (3): 510–526.

Summerour, C. W. 1953. "A Woodcock Nest Near Auburn." *Alabama Birdlife* 1: 10.

Tappe, P. A., R. M. Whiting, Jr., and R. R. George. 1989. "Singing-Ground Surveys for Woodcock in East Texas." *Wildlife Society Bulletin* 17 (1): 36–40.

Tappley, W. G. 2000. *Upland Days: 50 Years of Bird Hunting in New England.* New York: Lyons Press.

Tappley, W. G. 2009. *Upland Autumn.* New York: Skyhorse Publishing.

Teale, E. W. 1974. *A Naturalist Buys an Old Farm.* Storrs, Conn.: Bibliopola Press.

Thogmartin, W. E., J. R. Sauer, and M. G. Knutson. 2007. "Modeling and Mapping Abundance of American Woodcock across the Midwestern and Northeastern United States." *Journal of Wildlife Management* 71 (2): 376–382.

Thomas, D. W., and T. G. Dilworth. 1980. "Variation in Peent Calls of American Woodcock." *Condor* 82 (3): 345–347.

Thompson III, F. R., and R. M. DeGraaf. 2001. "Conservation Approaches to Woody, Early Successional Communities in the Eastern United States." *Wildlife Society Bulletin* 29 (2): 483–494.

Tordoff, H. B. 1984. "Do Woodcock Carry Their Young?" *Loon* 56 (Summer): 81–82.

Trani, M. K., R. T. Brooks, T. L. Schmidt, V. A. Rudis, and C. M. Gabbard. 2001. "Patterns and Trends of Early Successional Forests in the Eastern United States." *Wildlife Society Bulletin* 29 (2): 413–424.

Vale, R. B. 1936. *Wing, Fur, and Shot: A Grassroots Guide to American Hunting.* New York: Stackpole Books.

Vance, J. M. 1981. *Upland Bird Hunting.* New York: E. P. Dutton.

Vashon, J. H., A. L. Meehan, J. F. Organ, W. J. Jakubas, C. R. McLaughlin, A. D. Vashon, and S. M. Crowley. 2008. "Diurnal Habitat Relationships of Canada Lynx in an Intensively Managed Private Forest Landscape in Northern Maine." *Journal of Wildlife Management* 72 (7): 1488–1496.

Villard, M. 1998. "On Forest-Interior Species, Edge Avoidance, Area Sensitivity, and Dogmas in Avian Conservation." *Auk* 115 (3): 801–805.

Vitz, A. C., and A. D. Rodewald. 2006. "Can Regenerating Clearcuts Benefit Mature-Forest Songbirds? An Examination of Post-Breeding Ecology." *Biological Conservation* 127: 477–486.

Vitz, A. C., and A. D. Rodewald. 2007. "Vegetative and Fruit Resources as Determinants of Habitat Use by Mature-Forest Birds during the Post-Breeding Period." *Auk* 124 (2): 494–507.

Walker, W. A., and M. K. Causey. 1982. "Breeding Activity of American Woodcock in Alabama." *Journal of Wildlife Management* 46 (4): 1054–1057.

Waterman, C. F. 1972. *Hunting Upland Birds.* Selma, Ala.: Countrysport Press.

Wenstrom, W. P. 1974. "Habitat Selection by Brood Rearing American Woodcock." In *Proceedings of the Fifth American Woodcock Workshop*, pp. 154–177. Athens: University of Georgia Center for Continuing Education.

Westerskov, K. 1950. "Methods for Determining the Age of Game Bird Eggs." *Journal of Wildlife Management* 14 (1): 56–67.

Wetherbee, D. K. 1959. "Egg Teeth and Hatched Shells of Various Bird Species." *Bird-Banding* 30 (2): 119–121.

Wetherbee, D. K., and L. M. Bartlett. 1962. "Egg Teeth and Shell Rupture of the American Woodcock." *Auk* 79 (1): 117.

Whiting, R. M., Jr., D. A. Haukos, and L. M. Smith. 2005. "Factors Affecting January Reproduction of American Woodcock in Texas." *Southeastern Naturalist* 4 (4): 639–646.

Wildlife Management Institute. 2009. "Best Management Practices for Woodcock and Associated Bird Species." Washington, D.C.: Wildlife Management Institute.

Wiley, E. N., and M. K. Causey. 1987. "Survival of American Woodcock Chicks in Alabama." *Journal of Wildlife Management* 51 (3): 583–586.

Williams, T. 2004. *Wild Moments*. North Adams, Mass.: Storey Publishing.

Wilson, A. 1839. *Wilson's American Ornithology, with Notes by Jardine*. New York: T. L. Magagnos and Co.

Wishart, R. A. 1977. "Some Features of Breeding and Migration of Woodcock in Southwestern Quebec." *Bird-Banding* 48 (4): 337–340.

Wishart, R. A., and J. R. Bider. 1976. "Habitat Preferences of Woodcock in Southwestern Quebec." *Journal of Wildlife Management* 40 (3): 523–531.

Wood, H. B. 1945. "A History of Bird Banding." *Auk* 62 (2): 256–265.

Woolner, F. 1974. *Timberdoodle! A Thorough, Practical Guide to the American Woodcock and to Woodcock Hunting*. New York: Crown.

Worth, C. B. 1974. "Body-Bobbing Woodcock." *Auk* 93 (2): 374–375.

Wright, B. S. 1960. "Woodcock Reproduction in DDT-Sprayed Areas of New Brunswick." *Journal of Wildlife Management* 24 (4): 419–420.

Wright, B. S. 1965. "Some Effects of Heptachlor and DDT on New Brunswick Woodcock." *Journal of Wildlife Management* 29 (1): 172–185.

Yahner, R. H. 2003. "Response of Bird Communities to Early Successional Habitat in a Managed Landscape." *Wilson Bulletin* 115 (3): 292–298.

Yahner, R. H. 2008. "Bird Response to a Managed Forest Landscape." *Wilson Journal of Ornithology* 120 (4): 897–900.

index

Booming from the Mists of Nowhere: The Story of the Greater Prairie-Chicken
     By Greg Hoch

The Butterflies of Iowa
     By Dennis Schlicht, John Downey, and Jeff Nekola

A Country So Full of Game: The Story of Wildlife in Iowa
     By James J. Dinsmore

The Ecology and Management of Prairies in the Central United States
     By Chris Helzer

Fifty Common Birds of the Upper Midwest
     Watercolors by Dana Gardner, text by Nancy Overcott

Fifty Uncommon Birds of the Upper Midwest
     Watercolors by Dana Gardner, text by Nancy Overcott

Forest and Shade Trees of Iowa
     By Peter van der Linden and Donald Farrar

Iowa Birdlife
     By Gladys Black

The Iowa Breeding Bird Atlas
     By Laura Spess Jackson, Carol A. Thompson, and James J. Dinsmore

Of Men and Marshes
     By Paul L. Errington

Of Wilderness and Wolves
     By Paul L. Errington

Raptors in Your Pocket: A Great Plains Guide
     By Dana Gardner

*The Tallgrass Prairie Center Guide to Prairie Restoration
in the Upper Midwest*
By Daryl Smith, Dave Williams, Greg Houseal, and Kirk Henderson

Up on the River: People and Wildlife of the Upper Mississippi
By John Madson

Where the Sky Began: Land of the Tallgrass Prairie
By John Madson

Wildflowers and Other Plants of Iowa Wetlands
By Sylvan Runkel and Dean Roosa

Wildflowers of Iowa Woodlands
By Sylvan Runkel and Alvin Bull

Wildflowers of the Tallgrass Prairie: The Upper Midwest
By Sylvan Runkel and Dean Roosa